The Inkas:
Last Stage of Stone Masonry Development in the Andes

Francesco Menotti

BAR International Series 735
1998

Published in 2016 by
BAR Publishing, Oxford

BAR International Series 735

The Inkas: Last Stage of Stone Masonry Development in the Andes

© F Menotti and the Publisher 1998

The author's moral rights under the 1988 UK Copyright,
Designs and Patents Act are hereby expressly asserted.

All rights reserved. No part of this work may be copied, reproduced, stored,
sold, distributed, scanned, saved in any form of digital format or transmitted
in any form digitally, without the written permission of the Publisher.

ISBN 9780860549253 paperback
ISBN 9781407350561 e-format
DOI https://doi.org/10.30861/9780860549253
A catalogue record for this book is available from the British Library

BAR Publishing is the trading name of British Archaeological Reports (Oxford) Ltd.
British Archaeological Reports was first incorporated in 1974 to publish the BAR
Series, International and British. In 1992 Hadrian Books Ltd became part of the BAR
group. This volume was originally published by Archaeopress in conjunction with
British Archaeological Reports (Oxford) Ltd / Hadrian Books Ltd, the Series principal
publisher, in 1998. This present volume is published by BAR Publishing, 2016.

BAR titles are available from:

	BAR Publishing
	122 Banbury Rd, Oxford, OX2 7BP, UK
EMAIL	info@barpublishing.com
PHONE	+44 (0)1865 310431
FAX	+44 (0)1865 316916
	www.barpublishing.com

Close-up of the perfectly fitted stone of the Inka Roq'e Palace (Cuzco)
(Picture: F. Menotti 1995)

To my parents

(Dedicato a miei genitori)

ABSTRACT

Although stone masonry is a major hallmark of Inka culture, archaeologically speaking, not much work has been done on this in the last few decades. Indeed, the best scientific researches on this topic have been carried out by scholars outside the archaeological discipline, such as Agurto Calvo and Protzen. The main objective of this work is to give the the reader an accurate and detailed account of Inka stone masonry from an archaeological perspective. After a generic but not superficial overview on the Inkas and their former *Tawantinsuyu* (or Empire of the Four Quarters), the topic is considered in its whole from stone block extraction to dressing techniques, tools employment and stone block transport. The study includes an examination of the Inkas' five most exploited quarries within the Cuzco region with a special emphasis on the Llama Pit of Rumiqolqa Quarry.
Unlike what was thought a couple of decades ago, the main styles of stone wall construction are now divided into five categories: rustic, cellular, encased, sedimentary and cyclopean. Structural features of a wall (e.g. block shape, joints, cross-section, profile and texture) are considered in relation to the five categories. The applications of these styles as well as their subdivisions are reflected in various kinds of Inka buildings around the Sacred Valley of Cuzco. As practical examples I chose six of the fairly well known Inka sites: Machu Picchu, Ollantaytambo, the agricultural terraces of Moray, the Royal Estate of Pisaq, Tambo Machay and Saqsaywaman. Cuzco itself, the capital of the Inka Empire, is considered separately for it is the best representation of the transformation from an Inka centre into a Colonial town. After almost five centuries in a seismic area which is highly at risk, contrary to what has happened to Spanish edifices, Inka walls have proven to be earthquake proof. Consequently, the final consideration is a possible empirical application of the study: understanding ancient building techniques and, applying them to modern structures for a safer future.

ACKNOWLEDGEMENTS

There are many people to whom am very grateful for providing me with invaluable assistance during the preparation of this work.

Firstly, I would like to express my sincere gratitude to Associate Professor John Campbell who provided me with encouragement, useful suggestions and above all, his remarkable knowledge without which this work could not have been completed.

I would also like to thank a few Peruvian scholars who helped me during the two months of fieldwork I spent in Peru in 1995: first of all, Professor Alberto Bueno of the Department of Archaeology of the San Marcos University at Lima for letting me consult his private library; secondly, the archaeologist in charge of the excavation "Proyecto Qori Kancha" Cuzco, Lic. Walter Zanabria, for being my guide around the Sacred Valley of Cuzco and showing me the latest achievements concerning Inka stone masonry studies; thirdly, Professor Raymundo Bépo who was of great help at the San Antonio Abad University of Cuzco.

I am also indebted to a good friend, Chris Davies, for sharing with me some creative ideas and giving me useful "insides".

A special thanks goes to Mrs Lyn Burrows, the administrative officer of JCU Department of Anthropology and Archaeology, who has the special gift of being able to deal with all sorts of human beings. Thanks Lyn, I will never forget your kindness!

Finally, intimate thanks are due to my parents for their support throughout all my studies.
(Un grazie in particolare ai miei genitori senza i quali quello che ho ottenuto finora nella vita non sarebbe stato possibile. GRAZIE!)

TABLE OF CONTENTS

Chapter 1: Introduction   Page

    1.1 Aim of the thesis   1

    1.2 Sources of information   1

    1.3 Structure of the thesis   2

Chapter 2: Origins and history of the Inkas

    2.1 Introduction   4

    2.2 The legendary origins of the Inkas   5

    2.3 The mythological Inka dynasty   6

    2.4 The historical dynasty of the Empire   9

    2.5 The Spanish invasion and their puppet Inkas   11

Chapter 3: Aspects of the Inka Empire organisational systems

    3.1 Introduction   13

    3.2 The geographical position of the Inka Empire   13

    3.3 Sociopolitical organisation   14

    3.4 The administrative system   15

    3.5 The Inkas: their religion and beliefs   17

    3.6 The agriculturally based Inka economy   19

    3.7 The Inka military organisation   20

    3.8 The Inka "highway" network   21

Chapter 4: Inka stone masonry

4.1 Introduction	23
4.2 Origins of Inka stone masonry	24
4.3 Inka quarries: stone block extraction	25
4.4 Inka stone masonry tools	30
4.5 Inka stone dressing techniques	31
4.6 Stone block transport and manipulation	32
4.7 The different types of Inka stone wall construction	34

Chapter 5: Five types of Inka wall structure as applied to various buildings around the Sacred Valley

5.1 Introduction	43
5.2 Machu Picchu	44
5.3 Ollantaytambo	48
5.4 The agricultural terraces of Moray	51
5.5 The Pisaq Royal Estate	53
5.6 Tambo Machay	57
5.7 Saqsaywaman	57

Chapter 6: Cuzco: from Inka control to European Colonial planning

6.1 Introduction	61
6.2 Cuzco's geographical location	61
6.3 Prehispanic Cuzco	63
6.4 Qori Kancha: the Temple of the Sun	65

6.5 Cuzco: the "puma city"	66
6.6 Inka walls in today's Cuzco	66
6.7 Inka stone masonry: perspectives for the future	70

Chapter 7: Summary and conclusions   72

References   75

Please note that 100 colour photographs of stone and walling detail with more general views of Inka sites are available to download from www.barpublishing.com/additional-downloads.html

LIST OF FIGURES

		Page
Figure 2.1	The Inka dynastic succession	4
Figure 2.2	The geographical setting of the legendary journey of the Inka ancestors from Pacaritambo to Cuzco	6
Figure 2.3	Manqo Qhapaq	7
Figure 2.4	Zinchi Roq'e	7
Figure 2.5	Lloq'e Yupanki	7
Figure 2.6	Mayta Qhapaq	8
Figure 2.7	Qhapaq Yupanki	8
Figure 2.8	Inka Roq'e	8
Figure 2.9	Yawar Waqaq	9
Figure 2.10	Wiraqocha Inka	9
Figure 2.11	Pachakuti Inka Yupanki	10
Figure 2.12	The extent of the Inka Empire showing territories annexed between 1438 and 1525	10
Figure 2.13	Thupa Inka Yupanki	10
Figure 2.14	Wayna Qhapaq	11
Figure 3.1	The Tawantinsuyu: the land of the four quarters	13
Figure 3.2	The Inka socio-hierarchical pyramid	14
Figure 3.3	The Quipu: the Inka recording device based on knots and colours	14
Figure 3.4	The Inka meaning of colours	16
Figure 3.5	Diagram of the Inka religious cosmology	17
Figure 3.6	The Coya Raymi festival	19
Figure 4.1	The five most important Inka quarries in the Sacred Valley	25
Figure 4.2	The Llama pit at Rumiqolqa Quarry	26
Figure 4.3	Protzen's experiment on Inka stone dressing techniques	33

Figure 4.4	Transport of stone blocks depicted by Poma de Ayala	34
Figure 4.5	Cross-sections of structure of Inka walls	35
Figure 4.6	Inka stone blocks: shape	36
Figure 4.7	Inka stone block joints	36
Figure 4.8	Profiles and surface textures of Inka walls	37
Figure 5.1	Map of the Sacred Valley of Cuzco	43
Figure 5.2	General plan of Machu Picchu	45
Figure 5.3	Plan of the Ollantaytambo complex	49
Figure 5.4	Plan of the agricultural terraces complex of Moray	52
Figure 5.5	Plan of Pisaq complex	55
Figure 5.6	Plan of Saqsaywaman	59
Figure 6.1	Possible location of the two main Inka squares of Prehispanic Cuzco	62
Figure 6.2	Haukaypata and Kusipata squares depicted by Poma de Ayala	63
Figure 6.3	Schematic diagram of the Zeque system	66
Figure 6.4	Cuzco: the "puma city"	67
Figure 6.5	Plan of the Qori Kancha Temple	70

Please note that 100 colour photographs of stone and walling detail with more general views of Inka sites are available to download from www.barpublishing.com/additional-downloads.html

LIST OF PLATES

	Page
Plate 3.1 The bedrock sculptured Intiwatana of Machu Picchu	18
Plate 3.2 Inka weaponry: star-headed maces	20
Plate 3.3 The drawbridge near Machu Picchu	21
Plate 4.1 Inka double-jambed door	23
Plate 4.2 Different stages of stone block production at the Llama Pit of Rumiqolqa Quarry	27
Plate 4.3 A 2 kg Inka hammerstone	30
Plate 4.4 Protuberances on a sedimentary type of Inka wall at Ollantaytambo	38
Plate 4.5 Rustic style of stone wall construction	39
Plate 4.6 Cellular style of stone wall construction	39
Plate 4.7 Encased style of stone wall construction	40
Plate 4.8 Sedimentary style of stone wall construction	41
Plate 4.9 Cyclopean style of stone wall construction	42
Plate 5.1 Machu Picchu	44
Plate 5.2 The Temple of the Three Windows (Machu Picchu)	46
Plate 5.3 The Torreon of Machu Picchu	47
Plate 5.4 The Royal Tomb of Machu Picchu	47
Plate 5.5 The Wall of the Six "Monoliths" at Ollantaytambo	48
Plate 5.6 The portal of the sacred sector of Ollantaytambo	50
Plate 5.7 A joint of the six "monoliths" of Ollantaytambo	50
Plate 5.8 The agricultural terraces of Moray	51
Plate 5.9 Drainage system of Moray agricultural complex	53

Plate 5.10	Flights of stairs set into the terrace walls of Moray	53
Plate 5.11	The Pisaq Royal Estate	54
Plate 5.12	The agricultural terraces of Pisaq	54
Plate 5.13	The sedimentary style of stone wall construction at Pisaq	56
Plate 5.14	Composed semi-worked stone blocks with central filling	56
Plate 5.15	The Baths of the Inka (Tambo Machay)	57
Plate 5.16	The zigzag walls of Saqsaywaman	58
Plate 5.17	The foundations of the Muyuc Marca tower of Saqsaywaman	60
Plate 6.1	Cuzco's Plaza de Armas	61
Plate 6.2	The Amarukancha wall	64
Plate 6.3	The Acllahuasi wall	64
Plate 6.4	The retaining walls of Kolkampata	64
Plate 6.5	Part of the Kora Kora Palace	68
Plate 6.6	Bas-reliefs on the Amarukancha wall	69
Plate 6.7	The convent of Santa Catalina and part of the Acllahuasi wall	69

LIST OF TABLES

Page

Table 4.1 Types of rocks, hardness and compression   31
resistance in kg per cm^2

GLOSSARY OF QUECHUA TERMS

Acila: Virgins of the Sun

Acilahuasi: Convent of the Virgins of the Sun

Antisuyu: North-eastern quarter of the Inka Empire

Apusquipay: Military commander-in-chief

Aymaray Quilla: The cultivation festival

Ayamarka: Feast of the Dead

Ayllu: Community group

Camayos: local chiefs

Capac Siquis: Festival of the sowing month (August)

Zeque: Imaginary lines radiating from Cuzco

Chahua-Huarquiz: Festival of the plouging month (July)

Chasquis: Runners or messengers

Chinchaysuyu: North-western quarter of the Inka Empire

Chungacamayoc: Military guardians

Collasuyu: South-eastern quarter of the Inka Empire

Cuntisuyu: South-western quarter of the Inka Empire

Coya Raymi: Festival of the Moon

Curaca: Chiefs with Inka privileges

Hanan: One of the two moieties of Cuzco (upper)

Hatun Pucuy and Pacha Pucuy : Festivals of the ripening months (February and March)

Huaca (Waqa): Sacred places or objects related to the Zeque system

Hurin: One of the two moieties of Cuzco (lower)

Intiwatana: Astronomical observatory

Inti Raymi: The feast of the Sun

Llacta: Inka urban centre

Mit'a: Public work service or "tax in labour" system

Mitimae: Colonists

Pachamama: The Mother Earth

Quechua: Inka language

Quipu: Recording device based on knots and colours

Quipucamayocs: Quipu readers

Sapa Inka: Supreme Inka

Sinchi: Regional chief

Suyu: quarters or regions

Tambo: Inn or shelter

Tawantinsuyu: Land of the four quarters

Tocricoc: Provincial governors

Uma Raymi: Festival of the watching of the crops (October)

Wiraqocha: The Creator

The Inkas: last stage of stone masonry development in the Andes

Chapter 1

Introduction

1.1 Aims of the work

Since the Spaniards invaded the Inka Empire in AD 1532, thousands of litres of ink have been "poured" out trying to describe all aspects of that previously "unknown" culture. During the last 100-120 years scientific researches in various disciplines have considered the Inkas from a number of perspectives. But for unknown reasons and in spite of being one of the milestones of Inka culture, their stone masonry has not been studied in much detail. When 19th century naturalists such as Alexander Von Humboldt, Eduardo de Riviero and Diego de Tschudi travelled extensively through Peru, gathering information on Inka sites, their written records were mostly purely descriptive: i.e. site location, what the site looked like and the materials it was made of. The first studies were carried out by the German Alphons Stubel in 1877, subsequently Max Uhle in 1897 and more recently John Rowe who was the first to work out a chronology for the Inka dynasty in the 1940s (Moseley 1992:18).

Inka culture has been studied from a large number of perspectives: political, economic, symbolic, religious and many others. However, when considering the matter of architecture, one can notice that the amount of scholarly work done diminishes remarkably, especially that done by professionally trained archaeologists. In fact, the best detailed scientific studies have been carried out by scholars who have little or no archaeological training. For example, the work done by the Peruvian architect Santiago Agurto Calvo (1987) in the late 1980s has given us an in-depth understanding of the construction qualities behind the often impressive Inka stone buildings. Similarly, the study carried out by another architect, the Swiss Jean-Pierre Protzen (1983, 1985, 1986, 1993), explores all the technical aspects which made it possible for the Inkas to obtain such a highly developed expression of stone masonry.

A primary goal of this study is to fill the gap in the archaeological literature and write a detailed account of Inka stone masonry, from stone block extraction methods to dressing techniques, including the five main styles of stone wall construction. In other words, this work emphasises what is not visible to untrained eyes and provides a better understanding of the subject, thus giving people the opportunity to appreciate more properly what one can find out, not only from an aesthetical point of view, but from a technical one as well. This practical knowledge, if proved useful, can eventually be applied empirically to "modern" architecture in Peru and certain other earthquake prone areas in order to overcome possible construction problems.

1.2 Sources of information

Writers of biographies and historians are often fortunate to have at their disposal documents compiled by those who are the object of their studies. Unfortunately, archaeologists do not have this advantage in recalling the life of the Inkas, for writing was unknown to them. Therefore, scholars are limited to consulting Spanish chronicles, combined with actual archaeological research.

In collecting information for my research I have considered, within the limits of my own possibilities and skills of course, the most relevant Inka literature written in the last four centuries. Starting with the Spanish chronicles of the 16th and 17th centuries, I have then taken into account the first scientific writings of the 19th century and eventually the professional archaeological works of this century.

Although the chronicles are useful sources of information, an important aspect which has to be borne in mind, in most cases is that they are biased. Garcilaso de la Vega (1961 [1604]) for instance, being the son of an Inka princess and a Spanish nobleman, praises the Inkas somewhat excessively, whereas Sarmiento de Gamboa (1960 [1572]) pours scorn upon them. The writings of Pedro Cieza de Leon (1967 [1553]) are very descriptive and sometimes very sympathetic to the Inkas, but like other chroniclers, his detailed account of the *Tawantinsuyu* (i.e. the Empire of the "Four Quarters") must be regarded as a historical reconstruction (Moseley 1992:12). Father Cobo, on the other hand, in spite of his religious background, was one of the most "objective" commentators and his writings are often used by scholars of various disciplines. Another chronicler of the 16th and 17th centuries was Guaman Poma de Ayala (1980 [1583-1615]), although his written accounts of Inka society were only rediscovered at the beginning of this century (Baudin 1962:27). Unfortunately, Guaman Poma de Ayala's work is also not fully reliable, although some of his drawings have helped archaeologists and historians identify specific sites around Cuzco.

Scholarly study of a more "modern" sort on the Inkas began in the 19th century. The first book about South American prehistory including some accounts of Inka culture was published by William H. Prescott in 1847 (Beck 1983: 4). A few decades later, Max Uhle excavated a few sites in Cuzco, realising that the area had already been occupied by other cultures long before the Inkas (Valencia Zegarra 1979:3-8). Tello (1937), the first native Peruvian archaeologist belonging to the "Peruvian School", made his contribution to the archaeology of the Inkas by carrying on with what had been started by Stübel and Uhle, although his main interest was the pre-Inka period. At the beginning of the 20th century the American historian Hiram Bingham (1930) discovered Machu Picchu by accident while he was looking for the ruins of Vilcabamba. This event accentuated the interest in the Inkas and systematic studies began virtually straightaway. A good example of the latter is the work carried out by Rowe (1945, 1946) which brought to light the mytho-historical dynastic succession of Inka rulers from *Manqo Qhapaq*, first legendary Inka (AD c 1250), to the time of the Spanish invasion (1532).

In the last 30 years, a number of scholars have dedicated themselves not only to studying the Inkas in general, but also to other various aspects of their society. Gasparini and Margolies (1980), for instance, have considered Inka architecture in a broad sense within the entire Inka Empire, whereas Kendall (1984, 1989) has concentrated her attention on stone masonry in the Sacred Valley.

Patterson (1991), Moseley (1992) and Bauer (1992), on the other hand, have studied the *Tawantinsuyu* from an administrative perspective taking into account social division, economy and military organisation. Other scholars such as Malpass (1993), Topic and Lange Topic (1993), Schreiber (1993) and Lynch (1993) have undertaken their research with a regional approach studying various peripheral areas of the former Inka Empire to see the effects of the impact of Inka expansion in relation to local socioeconomic organisation.

In my research I have, of course, not only taken into account English-speaking scholars, but also South American and in particular Peruvian archaeologists such as Lumbreras (1974), Ravines (1978), Bonavia (1991) and Matos Mendieta (1994), who in spite of having been influenced by North American archaeological trends, have not been able to fill the gap completely between scientific archaeology and historical inquiry (Burger 1989:37). As mentioned in the first section, in order to construct a more global approach to the main topic, I will take advantage of the multidisciplinary nature of archaeology and consider works of scholars whose fields of studies might or might not have direct connections with the archaeological discipline. Within the social sciences for instance, I will cite historians such as Morris (1985), Rostoworoski (1988), Zuidema (1982, 1990), Espinoza Soriano (1990) and Patterson (1992) whereas outside the social science context, I am going to refer to scholars, such as the architects Agurto Calvo (1987) and Protzen (1983, 1985, 1986, 1993), and the mathematician Palomino Diaz (1994). Finally, the work is also supported by the material (pictures and maps) and information I personally collected during two months of fieldwork in Peru in 1995.

1.3 Structure of the research

After the present introductory chapter, Chapter 2 gives the reader a general background on the origins and history of the Inkas. With the help of Rowe (1945), Kendall (1989) and some well known chroniclers we will follow the development of the Inka Empire and its rulers' relative dynastic succession from the mythological

The Inkas: last stage of stone masonry development in the Andes

figure of *Manqo Qhapaq* ("first Inka") to the great *Pachakuti* (9th Inka and first historical ruler). We are also going to look at the post-Spanish-invasion period with its last so-called puppet Inkas nominated by Pizarro. Chapter 3 concentrates on various aspects of the *Tawantinsuyu* structural organisation. The sociopolitical hierarchical pyramid is compared with the administrative system, and it is shown that as far as the ranking division is concerned these two entities are very similarly structured. An important aspect of Inka society was religion with the two principal deities being The Sun and Wiraqocha, who gained more importance after *Pachakuti* saw him in one of his dreams. Religion, like many other activities, was deeply interwoven with the sociopolitical organisation. In fact, this is clearly shown in Zuidema's (1964, 1983) study concerning the *Zeque*, or imaginary lines, radiating from Cuzco and connecting a large number of *Huaca* (sacred objects or places) scattered around the city. According to Hyslop (1990: 220-221) a specific combination of *Zeque-Huacas* would have influenced the Cuzco class division as well as religious activity. Inka economy was essentially based on intensive agriculture, and in particular, on the cultivation of maize, tubers and beans. Evidence for this food production system can be seen throughout the former Inka Empire in the form of well made agricultural terraces which were provided with a remarkable irrigation system.

Most of the *Tawantinsuyu* (Inka Empire) was created within a span of a century. What made possible such a sudden and determined expansion was a well organised Army. Like other organisations, the Inka military society was strictly hierarchical. The Inka Army was formed in part with professional soldiers and in part with civilians temporarily recruited through the *mit'a* tax payment system (Espinoza Soriano 1990:360). The efficiency of this conquering force was also achieved because of the excellent road network which, at the heyday of the Empire, covered some 30,000 km in a total surface area of over more than 3,000,000 sq. km (Tamayo Herrera 1994:198). In Chapter 4 I will come to the main focus of my work: Inka stone masonry. After the first two sections which introduce the topic in general terms, giving a background about the origins of Inka stone masonry, the various processes of stone block extraction, stone block dressing, the usage of tools and transport techniques are discussed. As an example, I will consider a renowned quarry of the Cuzco region: the Rumiqolqa Quarry. Finally, our attention will be focused on the five principal styles of stone wall construction: rustic, cellular, encased, sedimentary and cyclopean. Within this description are also included wall cross-sections, block shapes, joints and surface textures.

Chapter 5 looks at how the technical systems of stone wall construction discussed in Chapter 4 were applied to various buildings around the Sacred Valley of Cuzco. Because of the large number of places which could be studied, I will limit my choice to six significant Inka sites which reflect the whole range of both common and royal Inka stone masonry: Machu Picchu, Ollantaytambo, the agricultural terraces of Moray, the Royal Estate of Pisaq, Tambo Machay (the "Inka Baths") and Saqsaywaman (the Inka ceremonial centre famous for its well fortified cyclopean walls just out of Cuzco). Amongst the above mentioned Inka sites I deliberately left out the city of Cuzco because it is considered separately in Chapter 6. It is interesting to see how this town underwent the transformation from being the capital of *Tawantinsuyu* to a Colonial agglomerate. Throughout Chapter 6 it is shown how the Spaniards dismantled most of the Cuzco's Inka buildings, but at the same time included some of them as foundations for their own Colonial edifices. This latter situation was originally done for purely aesthetic reasons, but it proved to have a major practical benefit during several earthquakes which have shown conclusively that the Inka walls were, in fact, earthquake proof. As a consequence, Inka stone masonry has become not only a tourist attraction, but also a technological innovation to be copied in new constructions.

The Inkas: last stage of stone masonry development in the Andes

Chapter 2

Origins and history of the Inkas

2.1 Introduction

When we talk about the origins of the Inka we often set chronological dates to distinguish different periods. One thing that has to be borne in mind is that the Inkas did not have any fixed points in time, they did not count their age, and they did not measure the duration of their acts in years. Consequently, as far as events in the past are concerned, they were just described as having happened *ñawpapacha* or "a long time ago" (Cobo 1979:252-253). Correlating with the Western calendar is almost impossible, and there is only one absolute date that can be related to the Inka Empire, viz. the 16th of November, 1532, the day the Spaniards, led by Francisco Pizarro, captured the last emperor *Ataw Wallpa*. Establishing the dates of events that took place before the arrival of the Spanish invaders is therefore a difficult and hazardous task (Bauer 1992:36).

In the light of these difficulties, the first attempt to establish an absolute chronology was made by Rowe (1945) in the early 1940s. He compared numerous descriptions of Inka dynastic successions found in the Spanish chronicles, and concluded that the ones provided in Cabello Balboa's work were the most plausible.

An important point that emerged from Rowe's study was the establishment of the approximate date of commencement of the Inka state just after *Yupanki* usurped the throne from *Wiraqocha Inka* on the eve of the *Chanca* invasion in AD 1438. Before this date Rowe (1945: 275) argued that the genealogies of the first eight Inka emperors, starting from *Manqo Qhapaq*, were highly implausible, since Cabello Balboa's dates would give those early rulers impossibly long reigns. Rather than using Cabello's successions, Rowe suggested that each Inka would have reigned for approximately twenty-five years. Therefore, according to his calculations the foundation of Cuzco by the mytho-historical *Manqo Qhapaq* would have occurred in around AD 1250. Rowe (1945:265) continued his well argued reconstruction along the lines that the dynastic list of Inka kings presented by Cabello Balboa is a mythical-historical continuum; the first eight Inka rulers are seen as primarily mythical while the last five are historical (see Fig. 2.1). Moreover, more recent studies done by Zuidema (1982) consider the above-mentioned Inka dynastic succession in a more interpretative way by arguing that the Spanish chroniclers presented pure mythical representation that cannot be read literally. According to Zuidema (1982: 173), Inka history, because it integrated religious, calendrical and ritual facts into one ideological system which was hierarchical in space and time, should

Ruler	Reign
Manco Capac	A.D. 1250: Founding of Cuzco
Sinchi Roca	Mythical
Lloque Yupanqui	Mythical
Mayta Capac	Mythical
Capac Yupanqui	Unknown
Inca Roca	Unknown
Yahuar Huacac	Unknown
Viracocha Inca	Until A.D. 1438: Chanca War
Pachacuti Inca Yupanqui	A.D. 1438–1471
Topa Inca Yupanqui	A.D. 1471–1493
Huayna Capac	A.D. 1493–1528
Huascar Inca	A.D. 1528–1532
Atahualpa	A.D. 1533–1533

Fig. 2.1. The Inka dynastic succession (after Rowe 1945:278)

actually be seen as mythological up to the time of the Spanish conquest.

It is important, Zuidema (1982:174) concludes: (1) that this hierarchical ideology is not confused with the Western linear conception of history imposed by the Spaniards; and (2) if a historical chronology of the Inka Empire up to Pizarro's invasion is to be formulated, this should be a task for archaeology. Similar concerns have been expressed by Morris (1988), Urton (1990) and Bauer (1992) who agree with the fact that the various accounts of the Inka kings written by the chroniclers may represent misinterpretations of the institution regarding Inka kinship. Even though a methodological framework for an independent chronology based on archaeological evidence has already been proposed by Rowe (1970) in the 1960s, Kendall (1989) in the 1980s, and more recently Bauer (1992) in the 1990s, very little scientific research has been undertaken in the last few decades. The widely and easily accepted Cuzco chronology written by the chroniclers has in effect hindered more costly and time-consuming scientific research.

Keeping in mind what has just been stated, in the next part of this chapter I will cite two of the most accepted mythological legends concerning the origins of the Inkas which were written by some of the famous chroniclers of the 16th and 17th centuries.

2.2 The legendary origins of the Inkas

As Urton (1990: 1) points out, one common theme in the mythological lore of ancient kingdoms around the world is the depiction of legendary ancestors as "strangers", people who appear from nowhere or from specific remote places and who undergo periods of primordial wandering before settling down in their new homeland. Of course they normally possess supernatural powers and advanced technologies, and they are destined to rule. The outline of the cycle of myths regarding the origins of the Inkas is not very different.

According to Sarmiento de Gamboa (1960 [1572]:224-228) four brothers and four sisters emerged from the central one of the three caves on a mountain called Tampu T'oqo near Pacaritambo, a place south of Cuzco. The principal figure of this group was *Manqo Qhapaq*, the man destined to become the first king of the Inkas. After organising the people living in the Pacaritambo area into ten groups called *Ayllu*, they set off in search of fertile land. Along their way they tested the soil a number of times with no success before arriving at a hill overlooking the valley of Cuzco. Recognising that the area was their long sought after home, the Inkas took immediate possession of the fertile valley (Classen 1993:39).

Similar versions of this legend are given by Father Bernabe Cobo (1979 [1653] Bk. 2, Ch.3: 103-105), a priest who travelled through the whole former Inka Empire in the 17th century. All three mythological stories have certain traits in common. For instance, one of them refers to two people (a man and a woman, possibly *Manqo Qhapaq* and his sister *Mama Oqllu*), who travelled from Lake Titikaka towards Cuzco in search of fertile land. They were given a rod of gold by their father the Sun and were told to thrust it into the ground from time to time. In the place where the golden rod would sink into the soil with the first blow, they should stop and make their settlement. After stopping at Pacaritambo, which in the Quechua language means inn or rest-house, they arrived in the valley of Cuzco where the staff of gold disappeared into the earth at the first attempt. They had found the right place to settle down.

There is not much difference between Cobo's description of the legend and that of Garcilaso de la Vega (1961 [1604] Bk. 1: 43-44). The Sun God Inti told his two children (a boy and a girl) to look for the promised land. They left Lake Titikaka, which was previously called Puquinacocha, and headed north towards Cuzco (Espinoza Soriano 1990:39). Also according to Garcilaso de la Vega the two children of the Sun stopped at the then mythological place called Pacaritambo. Finally upon arriving at the hill of Wanakauri, just before the Cuzco valley, they tried to thrust the golden rod into the ground. Not only did it sink at the first blow, but it disappeared under the earth's surface completely. The Inka settlement in the Cuzco region was thus about to begin.

The Inkas: last stage of stone masonry development in the Andes

According to the chroniclers, it is in this mystical and legendary atmosphere that *Manqo Qhapaq* started what was to become the biggest and for a time most successful empire of the New World: the Inka Empire. Of course, as is stated in the introduction, these legendary tales cannot be proven. But most of them reflect a peculiarity of the Inka mythology; they all have places in common which are real, and therefore they can be at least partly empirically studied. On this basis scholars such as Espinoza Soriano (1990) have been able to draw maps showing the mythical journey accomplished by *Manqo Qhapaq* and his brothers and sisters (see Fig.2.2). Furthermore, because of their real existence, these places are available for specific archaeological research projects like the ones carried out by Kendall (1989), Bauer (1992) and Malpass (1993).

2.3 The mythological Inka dynasty

As previously mentioned, information about the first eight Inka rulers is hazy and in many cases inconsistent. Rowe (1945: 275-283) considers the year 1250 a reasonable starting point for the first ruler and he believes that the period up to *Wiraqocha* represents a local development which corresponds archaeologically to the early Inka culture. In the following descriptions I will give the Quechua name of the ruler in modern Quechua spelling first and the Spanish-influenced version in parentheses immediately after. I have normally tried to show respect for the Quechua speakers by using their preferred spelling rather than the more widely established Spanish spelling. Quechua is a living language and was the language of the Inkas.

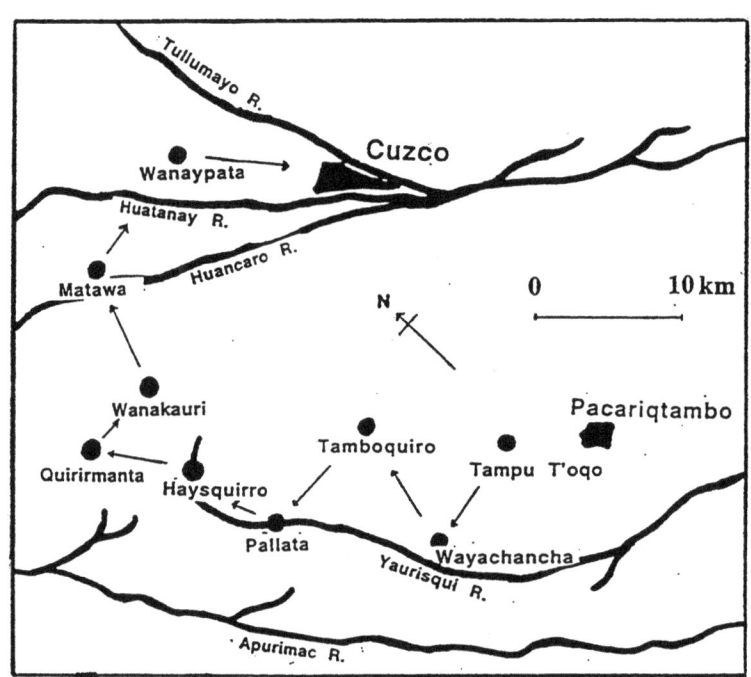

Fig. 2.2. The geographical setting of the legendary journey of the Inka ancestors from Pacaritambo to Cuzco (modified from Espinoza Soriano 1990:40)

In the next two sections of this chapter I will proceed the description of the Inka dynasty by listing the 13 Inka kings shown in Figure 2.1 and, with the help of both 16th/17th centuries' written records and more recent scientific works, we will follow step by step the development and expansion of the Inka state.

Manqo Qhapaq (Manco Capac)

According to Garcilaso de la Vega (1961 [1604] Bk. 1:48-52) *Manqo Qhapaq*, the first of the Inka kings, subjugated numerous villages, started to divide the Inka territory into four quarters and taught people how to worship the Sun God. He was loved and respected by his subjects because he gave

The Inkas: last stage of stone masonry development in the Andes

Fig. 2.3. Drawing of *Manqo Qhapaq* by Poma de Ayala [1615] (after Brundage 1985:28)

Zinchi Roq'e, Manqo Qhapaq's son, became the second king of the Inka dynasty at the age of twenty after the death of his father (Cobo 1979 [1653] Bk. 11:113). *Zinchi Roq'e* was a very influential man. His name means "the brave", but like his father he was generally a peaceful ruler. He travelled through the small kingdoms and gained the admiration of all his subjects (Angles Vargas 1988: 136). According to Garcilaso de la Vega (1961 [1604] Bk. 2:65), before *Zinchi Roq'e* died he passed on the throne to his legitimate son, *Lloq'e Yupanki*, born of his sister-bride *Mama Kura*.

Lloq'e Yupanki (Lloque Yupanqui)

them a "human condition" taking them away from a status of "barbarism". He was a very religious person and it was because of him that religion became part of the Inka sociopolitical life (Cobo 1979 [1653]:110). According to the legend, it was *Manqo Qhapaq* who started to build the magnificent temple of the Sun, called Qori Kancha, in Cuzco. After a long and successful reign, having reached old age, *Manqo Qhapaq* passed the command to his eldest son, *Zinchi Roq'e*. When he died his body was mummified and worshipped during religious rituals and agricultural feasts (Espinoza Soriano 1990:48).

Zinchi Roq'e (Sinchi Roca)

Fig. 2.4. Drawing of *Zinchi Roq'e* by Poma de Ayala [1615] (after Espinoza Soriano 1990:50)

Fig. 2.5. Drawing of *Lloq'e Yupanki* by Poma de Ayala [1615] (after Espinoza Soriano 1990:53)

According to Kendall (1989:25) *Lloq'e Yupanki*, as well as his father, *Zinchi Roq'e*, were peace-loving emperors. This statement is in contradiction with that of Garcilaso de la Vega (1961 [1604] Bk. 2: 66-68) who states that *Lloq'e Yupanki* did conquer a number of villages and other small groups in the area surrounding Cuzco. Divergent opinions also exist between Cobo (1979 [1653] Bk. 11:116), who argues that *Lloq'e Yupanki* married a woman from outside Cuzco, and Garcilaso de la Vega who states that he married his sister. One matter that seems to be of general agreement is that after his death his place was taken by his fourth son, *Mayta Qhapaq* (Espinoza Soriano 1990:52).

The Inkas: last stage of stone masonry development in the Andes

Mayta Qhapaq (Mayta Capac)

Fig. 2.6. Drawing of *Mayta Qhapaq* by Poma de Ayala [1615] (after Espinoza Soriano 1990:54)

Mayta Qhapaq, unlike his predecessors, had an aggressive character (Valdiva Carrasco 1988:28). Cieza de Leon (1967 [1553]) and Sarmiento de Gamboa (1960 [1572]) state that there were some quarrels between the Inkas and their neighbours. *Mayta Qhapaq* invaded those small communities, looted them and imposed some sort of tribute payment on them. According to Cobo (1979 [1653]) *Mayta Qhapaq* had two sons, *Tarco Huaman* and *Qhapaq Yupanki*. Later the latter became the fifth Inka.

Qhapaq Yupanki (Capac Yupanqui)

Fig. 2.7. Drawing of *Qhapaq Yupanki* by Poma de Ayala [1615] (after Espinoza Soriano 1990:57)

Following the chronicler Anello Oliva and disagreeing with Cobo's statement, Valdivia Carrasco (1988:30) argues that *Qhapaq Yupanki* was not *Mayta Qhapaq*'s son but another Inka who took power after a coup d'etat. *Qhapaq Yupanki* was the first Inka ruler who conquered lands outside the immediate Cuzco district, although these territories were only 20 km or so away. Even though *Qhapaq Yupanki* had many illegitimate affairs, he really loved his legitimate wife, with whom he had two sons, one of them, *Inka Roq'e*, becoming the sixth Inka emperor (Cobo 1979 [1653] Bk. 11: 122).

Inka Roq'e (Inca Roca)

Fig. 2.8. Drawing of *Inka Roq'e* by Poma de Ayala [1615] (after Espinoza Soriano 1990:58)

According to Baudin (1962:53), the sixth Inka king initiated the tribal group called Hanan, one of the two moieties of Cuzco. Apart from subjugating some local groups in the Cuzco area, *Inka Roq'e* did not attempt any major conquest. On the other hand, he was the first emperor who successfully fought against the Inkas' arch enemy, the *Chanca*. *Inka Roq'e* married *Mikay Qoqa* and their first son, *Yawar Waqaq*, succeeded him to the throne (Cieza de Leon 1967 [1553]).

Yawar Waqaq (Yahuar Huacac)

Yawar Waqaq ("he who weeps blood") had been given this name because he was once

The Inkas: last stage of stone masonry development in the Andes

Fig. 2.9. Drawing of *Yawar Waqaq* by Poma de Ayala [1615] (after Espinoza Soriano 1990:63)

Fig. 2.10. Drawing of *Wiraqocha Inka* by Poma de Ayala [1615] (after Espinoza Soriano 1990:64)

seen weeping blood out of grief. His vassals had a very poor opinion of him and he was also regarded as being a coward. However, unlike his father, *Yawar Waqaq*'s son developed a very aggressive character and therefore was sent away from the village. Later on, when the Inkas were once more confronted by the *Chancas*, the young prince took lead of the army, and in the name of *Wiraqocha* ("the creator"), he defeated the enemy and became the eighth Inka (Cobo 1979 [1653] Bk. 11:127-129).

Wiraqocha Inka (Viracocha Inca)

When the young prince became king, he was given the name of *Wiraqocha* because one night in a dream he saw the supreme divinity of *Wiraqocha*. The person who appeared in front of him was a man with a long and thick beard. This is the reason why the Inkas called the Spaniards *Wiraqocha* when they saw the bearded people of Pizarro's for the first time (Garcilaso de la Vega 1961 [1604] Bk. 5:172). According to Mason (1975:118) *Wiraqocha* was the first true imperialist, the first emperor who planned permanent conquest over foreign non-Inka people. As already mentioned, he defeated the *Chanca*, a neighbouring population with whom the Inkas had always been at war, forever gaining the respect of all his subjects. *Wiraqocha* had three royal sons.

Even though he nominated his favourite one, *Urqon*, as his successor, the third son *Kusi Inka Yupanki*, later known as *Pachakuti*, took power, becoming the ninth Inka and the first historical emperor.

2.4 The historical dynasty of the Empire

The death of *Wiraqocha Inka* marks the end of the legendary era of Inka history. Up until this time, the various chroniclers are sometimes in great disagreement. Only a few details can be given as likely to be incontrovertible and in most cases the truth is beyond confirmation (i.e. from a scientific point of view hypotheses can not be falsified). With the event of the next emperor, *Pachakuti*, all the major and more reliable authorities are in virtual agreement.

Pachakuti Inka Yupanki
(Pachacuti Inca Yupanqui) **(AD1438-1471)**

The historical Inka Empire began with the reign of *Pachakuti Inka* in 1438 and ended with the Spanish conquest in 1532. *Pachakuti* was a very powerful emperor.

The Inkas: last stage of stone masonry development in the Andes

Fig. 2.11. Drawing of *Pachakuti Inka* by Poma de Ayala [1615] (after Espinoza Soriano 1990:79)

He consolidated previous conquests and added new lands to the Empire, even though according to Rowe (1945:270-272) the main campaign of expansion was carried out by his son *Thupa Inka* between AD 1463 and 1471 (see Fig. 2.12).

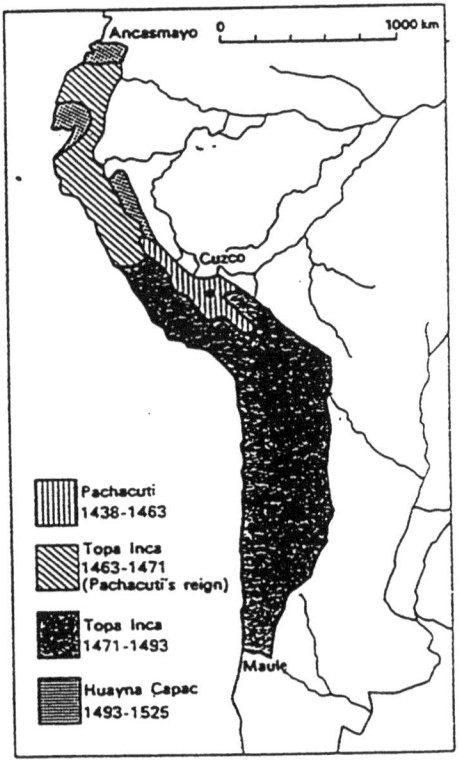

Fig. 2.12. The extent of the Inka Empire (after Kendall 1989:27)

Pachakuti was an excellent statesman. He organised the Empire with harmony, injected order and reason in everything, made the religious cult more extensive and established new festivals and ceremonies. More importantly, he organised an economic reform based on labour taxation and improved the road network substantially throughout the whole Empire (Cobo 1979 [1653] Bk. 11:133). *Pachakuti* had four children by his principal wives and one of them, *Thupa Inka Yupanki*, became the tenth king and emperor of the Inka Empire (Joyce 1969:93).

Thupa Inka Yupanki
(Topa Inca Yupanqui) **(AD 1471-1493)**

Fig. 2.13. Drawing of *Thupa Inka* by Poma de Ayala [1615] (after Espinoza Soriano 1990:94)

Thupa Inka was the greatest conqueror in the history of the Inka Empire. His successful campaigns added almost 80% of the territories to the Empire existing at the time of *Pachakuti*. He conquered and annexed almost the entire northern part of the Empire when his father was still the Inka ruler between AD 1463 and 1471, as was just pointed out above. As soon as he became the new emperor in 1471 he expanded his military conquests towards the south, reaching first Bolivia and then Chile as far as the River Maule (see Fig. 2.12).

He subsequently divided the conquered lands into four *Suyu* (quarters or regions), giving the empire the name of *Tawantinsuyu* (the land of the Four Quarters) (Espinoza Soriano 1990:97). According to Garcilaso de la Vega (1961 [1604] Bk. 8:309) *Thupa Inka* had six legitimate sons, but the honour of becoming the eleventh Inka was taken by *Wayna Qhapaq,* his eldest and favourite.

The Inkas: last stage of stone masonry development in the Andes

Wayna Qhapaq (Huayna Capac) (AD 1493-1528)

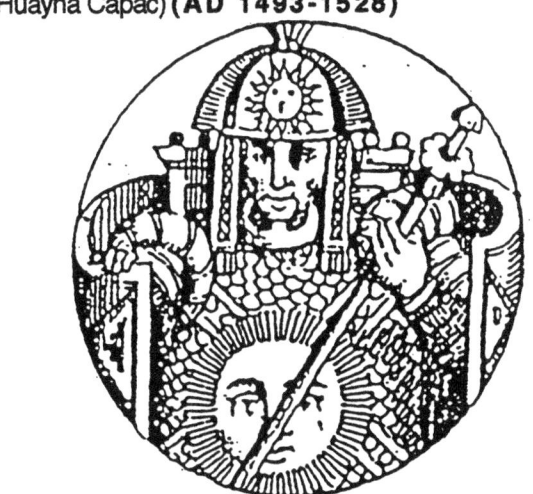

Fig. 2.14. Drawing of *Wayna Qhapaq* by Poma de Ayala [1615] (after Espinoza Soriano 1990:100)

After the death of his father in 1493, *Wayna Qhapaq* became the new ruler of the Inka Empire. By that time the Empire was almost complete, the only lands he added being the ones of Chachapayas, Moyobamba and some territories to the north of Quito, near today's border between Ecuador and Colombia (Kendall 1989:28). According to Cobo (1979 [1653] Bk. 11:163) *Wayna Qhapaq* had two sons, *Washkar* and *Ataw Wallpa*, who in the last five years of the Empire before the Spanish invasion, started a civil war which led the Inka state towards instability and its ultimate destruction.

Washkar (Huascar) and *Ataw Wallpa* (Atahuallpa): the last fully independent Inkas (AD 1528-1533)

Wayna Qhapaq's sons *Washkar* (previously called *Thupa Kusi Gualpa*) and *Ataw Wallpa* were the last rulers of the mighty Inka Empire (Cobo 1979 [1653] Bk.11: 163). When *Wayna Qhapaq* died *Washkar* was crowned king, but the northern part of the *Tawantinsuyu* was given to his brother, *Ataw Wallpa* (Sarmiento De Gamboa 1960 [1572]:256). *Ataw Wallpa* had excellent military training and, following his aggressive nature, took advantage of it by trying to overtake his brother's Empire. After a few attempts he managed to capture *Washkar*, but his success lasted but a very short time. In fact, at about the same time the Spaniards, led by Pizarro, arrived in Peru and were about to invade the already unstable Empire (Burland 1976:217).

2.5 The Spanish invasion and their puppet Inkas

November 16th, 1532 is the unfortunate date which signals the end of the most powerful indigenous state of the Southern Hemisphere: the Inka Empire. On this day Francisco Pizarro and his men arrived in Cajamarca and captured the emperor *Ataw Wallpa* (Marrin 1989:99). At the beginning the Spaniards asked for an enormous ransom for his liberation. However, after having tried him, they found him guilty of stirring up the native population and various other supposed crimes. Consequently, *Ataw Wallpa* was garrotted to death the following year on 29th, August 1533 (Guidoni and Magni 1977: 119).

After the execution of the last Inka emperor, Pizarro selected a number of "puppet" Inkas to represent the Empire. The first one was known as *Thupa Wallpa*, a brother of *Washkar*'s. Soon after, *Thupa Wallpa* was poisoned by *Ataw Wallpa*'s followers, and Pizarro had to nominate another puppet Inka. The choice of *Manqo Inka* was not a happy one from the Spanish point of view, as later on in 1536, he rebelled against the Spanish Conquistadors and established the Inka jungle state of Vilcabamba. While *Manqo Inka* maintained his neo-Inka state, a new puppet Inka called *Pawllu Thupa* was invested by the Spanish. *Pawllu Thupa* was followed by *Carlos Inka*, who "ruled" Cuzco for 23 years from 1549 to 1572. Meanwhile, in the Vilcabamba state, *Manqo Inka* was succeeded by three other rulers: *Sayri Thupa Inka* (1545-1558), *Titu Kusi* (1558-1571) and *Thupa Amaru* (1571-1572). The rule of the last was cut short when he was captured by the Spaniards and executed in 1572 (McIntyre 1975:199). This effectively ended the Vilcabamba state. Some high altitude centres such as Machu Picchu may have carried on longer, though this remains to be proved.

As we have seen so far, the Inka Empire developed from uncertain events shrouded in mythological legends. Thanks to scholarly works, such as that of Rowe (1945), it has been possible to give the different events at least somewhat arbitrary dates. However, a problem which arises thanks to this arbitrarily fixing of what are still imaginary chronological dates is that events which are related to determined fixed points in time do not necessarily explain the facts themselves. A good example is the historical beginning of the Inka Empire as determined by the Inka victory over the *Chanca* under the leadership of *Pachakuti*. That makes it look as if it was the *Chanca* war which started the Inka Empire. In reality, as shown in the studies carried out by Julien (1983), Kendall (1989), Bonavia (1991) and Bauer (1992), the development of the Inka state seems to be a reflection of various events and a combination of struggles between different ethnic groups which eventually resulted in the prevalence of the Inkas over the others. This perspective of continuity and acculturation explains the similarities found between the two pre-Inka Andean civilisations, the *Wari* (Huari) and the *Killke*, and the Inka culture itself (Bonavia 1991: 533-538).

Now that we know more or less how the Inka Empire came into being, in the next chapter we will consider it from an administrative point of view. We will see that the secret of its apparently "perfect" organisation was its capacity for integrating local traditions with the central administration of Cuzco. In other words, while imposing their administrative rules on the newly conquered lands, the Inkas normally maintained most forms of local socio-economic organisation, thus avoiding dissatisfaction and rebellion, say of the sort which occurred from time to time in the Roman Empire of the Old World and the Aztec Empire of Mexico (Grosboll 1993: 45; see also Fulford 1992; Kallet-Marx 1995; Smith and Berdan 1992).

The Inkas: last stage of stone masonry development in the Andes

Chapter 3

Aspects of the Inka Empire organisational systems

3.1 Introduction

Now that a general background on the origins and development of the Inka Empire has been presented, in this chapter I will consider the diverse aspects of its organisation. The highly hierarchical division of the society influenced all aspects of social life from political to military and even religious. However, no matter how elevated a person's social status might have been, nobody was above the *Sapa Inka* (i.e. the emperor). The secret of the nearly "perfect" Inka administrative organisation was its rigidity. The legal system was very strict and capital punishment was often applied. To many of us today it seems unbelievable that such an efficient society did not develop any writing system. The only recording device used was the so-called *Quipu,* a series of knotted and coloured cords (see section 3.4 below).

Religion and beliefs played an important role in the Inkas' life. Although *Wiraqocha* was reputed to be the "creator", the most commonly worshipped god was the Sun. Religious life was not an independent part of the social system, but it was deeply integrated in both the political and administrative organisations.

The Inkas were also very skilled astronomers, and according to Diaz (1994: 79) they were able to understand solar and lunar movements through complex mathematical calculations. The economy was mainly based on agriculture, even though other activities such as herding, fishing, weaving, pottery and metal working were definitely part of the total economic production system. Like religion, agricultural activity was related to a number of festivals which were celebrated throughout the year according to the different agricultural seasons. There was no actual market or any sort of currency within the Inka Empire. Taxes were paid rather through the *mit'a,* which consisted of asking people to work for the state for a certain period of time.

The spread of the *Tawantinsuyu* (the Inka state or land of the "Four Quarters") started with the ninth Inka ruler, *Pachakuti*. The successful campaign of expansion was particularly due to the efficient Inka Army, formed in part of professional soldiers and in part of common people recruited through the *mit'a* tax payment system. Finally, the mobility of this conquering force was essential. The Inkas, therefore, created a remarkable road network, thus assuring an excellent linkage between their heartland (the Cuzco Valley) and the peripheral regions of the Empire.

3.2 The geographical position of the Inka Empire

As pointed out earlier, the Inka Empire was divided into four parts called *Suyu* (quarters) and this is the reason why the state was also called *Tawantinsuyu* (land of the "Four Quarters"). Each quarter had a specific name: the North-western part was called *Chinchasuyu*, the North-eastern *Antisuyu*, the South-western *Contisuyu* and the South-eastern *Collasuyu* (see Fig. 3.1; see also Beck 1983:32).

Fig.3.1. The *Tawantinsuyu*: the land of the four quarters (after Moseley 1992: 26)

The position of these four quarters, (themselves sub-divided into small provinces which corresponded to the pre-

Inka native boundaries) was reflected in the plan of Cuzco which in the Quechua language means "navel". In fact, according to Inka cosmology, Cuzco was ideologically located in the "centre of the world". In terms of physical boundaries the Inka Empire was enormous. At the time of the Spanish invasion it stretched from Ecuador down to Central Chile with a coastal distance of about 4,500 km, and an area close to 3 million sq. Km (Tamayo Herrera 1994:198).

3.3 Sociopolitical organisation

The autocratic and centralised system which formed the basis of Inka society was typically pyramidal (see Fig. 3.2) with at its summit a sole and absolute authority in the person of the *Sapa Inka*, the supreme Inka, who was thought to be the descendent of the Sun God, as mentioned in Chapter 2.

Fig. 3.2. The Inka socio-hierarchical pyramid (after Kendall 1989:6)

This social position could only be inherited, and most of the time in order to maintain the "pure" blood the Inka emperor would marry his own sister. Just below the "Inkas by blood" were the "Inkas by privilege". This latter social class was founded by *Pachakuti*, who for administrative purposes gave this "new" status to those who showed any administrative ability for enforcing the Inka government policies (Espinoza Soriano 1990:276). Descending the "social pyramid" we find first the *Apus* or the Inka's council of the four prefects or quarters (the *Suyus*), and then the so-called *Tocrico*, or the provincial Governors. When the Inkas took over new lands, they usually let the native chiefs or kings (the *Sinchi*) of those lands rule the conquered territories. These ruling chiefs and kings were also called *Curaca*, and they had some privileges from the Inkas. As shown in Fig. 3.2, they had different names according to the number of people they ruled; the lowest (i.e. 100) people ruled were called *Pacha Curaca* and the highest (10,000) people ruled were the *Huanu Curaca* (Kendall 1989:57). Just above the ordinary "Commoners" or subject people there were the *Camayo*, that is some sort of "foremen" who used to lead between 10 and 50 householders. The subjects were the taxpayers whose work provided the produce and goods on which administration was supported. As pointed out above, the actual tax paid was calculated in work performed rather than goods, in other words, their tribute was their pure labour.

Within the Inka society there was one characteristic feature which was always present. This is the *Ayllu*, a general Inka term to describe the basic Andean social and organisational group (Bushnell 1963: 131). An *Ayllu* had a founding ancestor and contained a number of lineages divided between two moieties (Moseley 1992:49). This bipartite division was a traditional pre-Inka concept of social organisation. As far as the capital city of Cuzco was concerned, most chroniclers, such as Guaman Poma de Ayala (1980) [1615], Cobo (1979) [1653] and Garcilaso de la Vega (1961) [1604], agree that the town itself was divided into *Hanan* (upper) and *Hurin* (lower) parts. Much speculation has been made in an attempt to find a plausible answer for the possible function of these two moieties. Zuidema (1990), for instance, argues that because *Pachakuti* was responsible for the reorganisation of Cuzco, those who were of primary kin to the emperor were put in the *Hanan* section, while his distant relatives or illegitimate brothers, would belong to *Hurin* Cuzco. This division might also have provided a framework for marriage relationships which were, apart from the Inka emperor, essentially exogamous (i.e. out of the *Ayllu*).

3.4 The administrative system

According to Garcilaso de la Vega (1961 [1604] Bk. 5:156) as soon as the Inka conquered new lands, he immediately sent the *mitimaes* (or colonists) to take care of the conquest. This and the employment of the *Curaca* (local chiefs) became the two most important points of Inka success in administration. By doing this the Inka spread his supremacy over the state but at the same time preserved the authority of the local leaders, thus avoiding rebellion and dissatisfaction. Strict adherence to this principle therefore explains why the different regions of the Empire kept, to a certain degree, their diverse characteristics (Farrington 1992:379-381). Furthermore, according to Millones (1992:204), even though the Inkas tried to impose their language (Quechua), local populations outside Cuzco continued to speak their own language and dialects with Quechua only being used for administrative matters. Another hallmark of the Inka administration was law enforcement. Prescott (1970:38-39) argues that the punishments were very severe. They were almost wholly related to actual criminal acts, although sometimes blasphemy against the Sun God and the emperor were also punished by death. The chronicler Cobo (1979 [1653] Bk.11 Ch.26:203) lists a number of crimes which were dealt with by imposing the death penalty: robbery, killing by treachery, killing by casting spells, adultery, seduction of an *Aclla* (or virgin of the Sun see below), amongst many others. Like social organisation, administration was very hierarchically based. Any decision was made by following the social pyramid mentioned in the previous section.

As already stressed above, a peculiarity of Inka culture is that it did not develop any sort of writing system or calendar, given that it did develop many of the other fully bureaucratic structures of a state society. The only recording device used was the so-called *Quipu* (see Fig. 3.3), which consisted

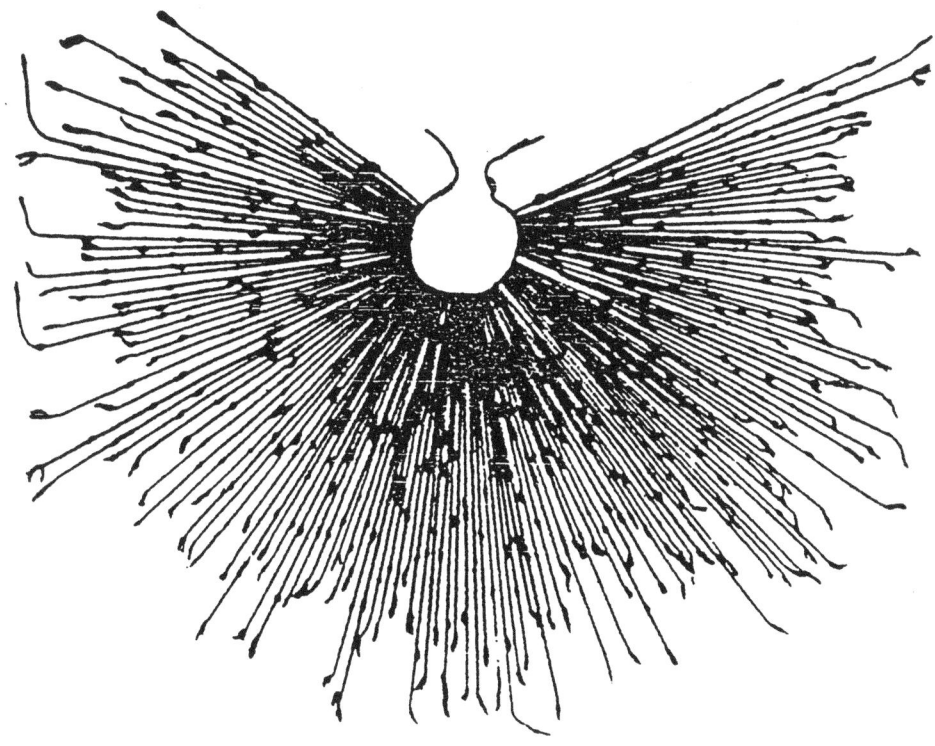

Fig. 3.3. The Quipu: the Inka recording device based on knots and colours (after Morris 1988:250)

of knotted and coloured cords. From a main cord hung coloured woollen threads. The position and number of knots tied on these threads provided numerical information based on a decimal system (Dobyns and Doughty 1976:51). Palomino Diaz (1994:70-73) has studied the meanings of colours within Inka culture (see Fig. 3.4) and he argues that the same system was adopted with the *Quipu*.

At the same time Diaz remains skeptical about the proper function and meanings of the knots. Only specialised people called *Quipucamayoc* were able to decipher or interpret the *Quipu*, and only men were entitled to hold this job. An interesting fact is that each *Quipucamayoc* had a personal *Quipu* which only he was able to understand (Stierlin 1984:190).

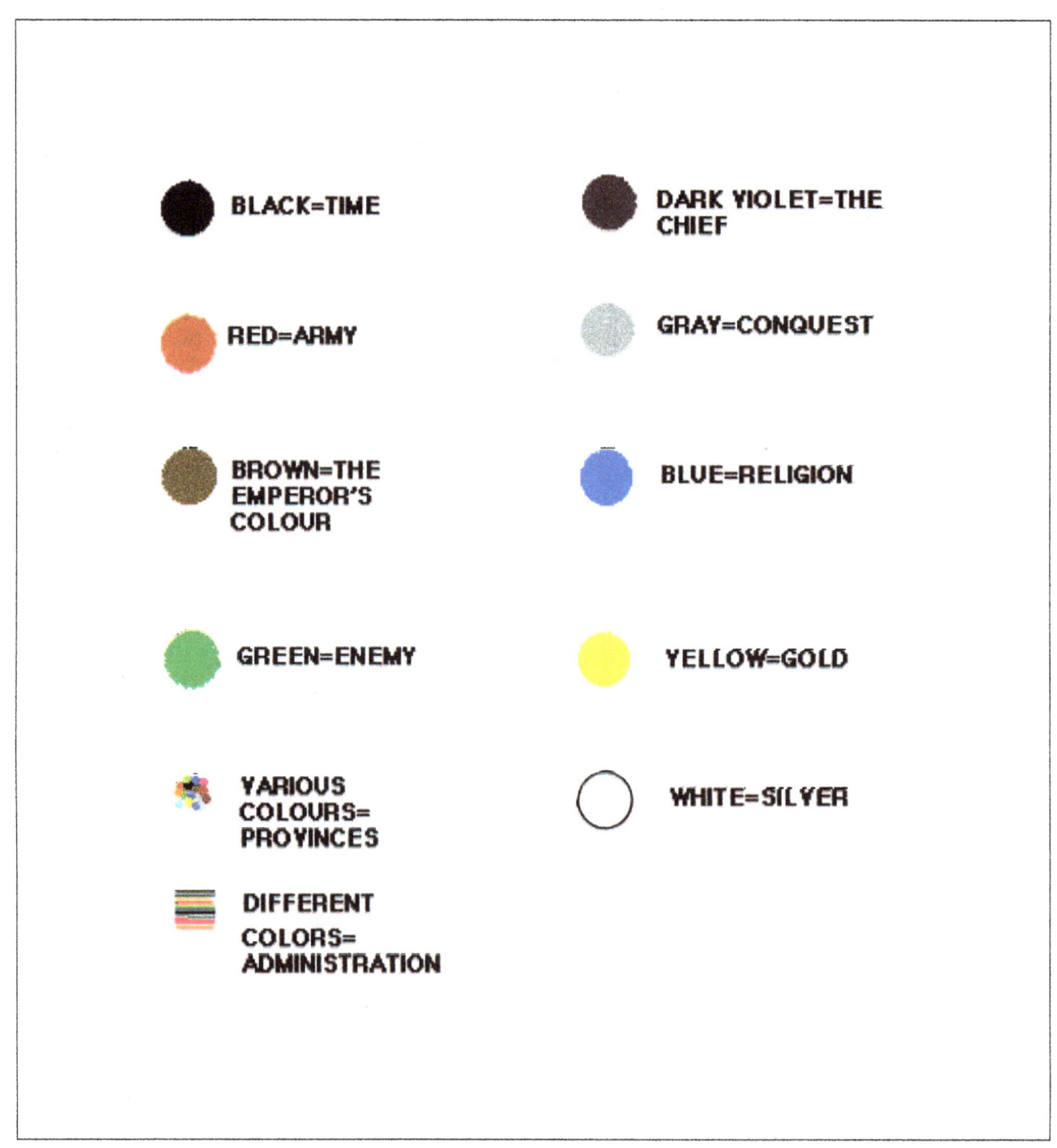

Fig. 3.4 The Inka meaning of colours (drawing F. Menotti 1996)

The Inkas: last stage of stone masonry development in the Andes

Within the hierarchical structure of the administrative force there were also the Army and the clergy. Even though the latter had an independent status, they had to conform with the emperor's wishes and therefore adhere to his administrative decisions. Outside the immediate hierarchical structure there were two other groups. The first one was the *Yana* or hereditary servants of the Inka and the Sun. According to Kendall (1989:62-63), they started as an underprivileged group, but by the time of the Spanish invasion they had become a class with a distinct position within the society. The second group, which was not directly involved with or influenced by the hierarchical pyramid of administration, was the *Aclla* or virgins of the Sun. This particular group of "chosen women" will be considered in greater detail in the next part of this chapter.

3.5 The Inkas: their religion and beliefs

Religion, superstition and legend were woven deep into the Inkas' life. Before *Pachakuti*, according to Cieza de Leon (1967 [1553]), the creator of all things was thought to have been the Sun. It is believed that Pachakuti, after a dream in which he saw *Wiraqocha* (the creator), drew a diagram (see Fig. 3.5) on the wall of the Temple of the Sun (*Qori Kancha*) describing Wiraqocha as both the father and the mother of the Sun and the Moon (Kendall 1989:182).

Although *Wiraqocha* was the "creator", the Sun was still the most important of the deities and it was regarded as the pivot of the Inkas' cosmology. In order to worship it, magnificent temples covered in gold were built, with the most famous one being the above mentioned *Qori Kancha* (place of gold) situated in Cuzco (Guidoni and Magni 1977:133).

In addition to the hierarchical ladder of priests and other clergy devoted to the Inka deities, as mentioned above there was another independent group called *Aclla* or virgins of the Sun (Von Hanstein 1971 [1925]:56). Every year a number of young girls who gave promise of exceptional beauty were chosen throughout the land by appointed officials and were taken to special convents called *Acllahuasi*. These "chosen women" served the priests in their functions and on particular occasions, especially critical war periods, the most attractive ones were sacrificed to the Sun God.

The Inkas did not only worship the Sun. They also had the cults of *Mama Quilla* (the Mother Moon), the stars and *Pachamama* (the Mother Earth), a female deity associated with agricultural activity (Kendall 1989:189). Besides the official Gods, the Inkas worshipped some holy sites or sacred shrines called *Huaca* (*Waka*). According to Beck (1983:49) Cuzco itself was a *Huaca* which was connected through "*the Zeque*" (imaginary lines radiating from the city) to numerous *Huaca* scattered around the Cuzco Valley. *Huaca* assumed different forms: buildings, little objects, places, caves, quarries, springs, hills and other natural features. Studies carried out by Zuidema (1964, 1983, 1990) and Williams Leon (1992) show that the principle of organisation found in the religious *Zeque* system could also have been fundamental to measurement of time and to the sociopolitical division of the Empire.

An important activity related to religion was astronomy. The Inkas were skilled astronomers, and they learnt how to combine the 12 lunar months with the 12 solar ones (Ziolkowski and Sadowski 1989:131). Related to this astronomical activity, many observatories were built around the Empire with the most famous ones being known as *Intiwatana*. These

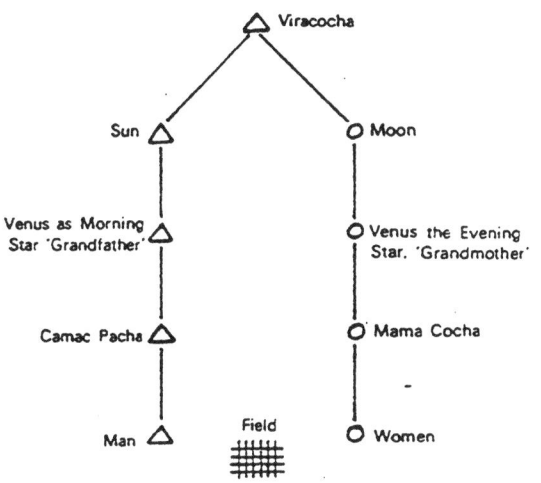

Fig.3.5. Diagram of the Inka religious cosmology (after Kendall 1989:184)

monuments which were sculptured in bedrock and had a specific function: to see the movement of the Sun. In fact, the word *Intiwatana* in Quechua means "where the sun is tied up". Like Maria Reiche (1989 [1968]) with her work on the Nazca lines, the mathematician Palomino Diaz (1994) has tried to study the function of the *Intiwatana* from a mathematical perspective. Diaz (1994:77-81) argues that these observatories were made by following complex calculations in order to understand the movements of the Sun. Each sculptured side had specific functions which have not been fully understood yet.

month of June. During this festival many sacrifices were made, such as the killing of llamas, the offering of the bodies of the *Aclla* (Virgins of the Sun) to the Sun and, according to some archaeological evidence, the burying of children alive with silver and golden vessels (Kendall 1989:198). Dancing and drinking accompanied the ceremony which normally lasted for several days.

There were other important feasts throughout the Inka year. For example, the *Coya Raymi* or the festival of the Moon, which was celebrated in September (see Fig. 3.6).

Plate 3.1. The bedrock sculptured *Intiwatana* of Machu Picchu (picture: F. Menotti 1995)

There are some good examples of *Intiwatana* all around the Sacred Valley, but the best known is the one in Machu Picchu (see Plate 3.1). As mentioned above, the Inkas divided the solar year into 12 months in which they used to perform various religious ceremonies and festivals. The most famous one was the *Inti Raymi,* the feast of the Sun, which is celebrated even today (albeit in a more mild manner) in the

The *Ayamarka* or feast of the Dead was held in November. The magnificent *Capac Raymi* festival was held in December and finally the great cultivation festival related to agriculture, the *Aymaray Quilla,* was celebrated in May. This series of rituals and celebrations were, voluntarily or involuntarily, also part of the Inka economic system on which state subsistence was based.

Fig. 3.6. Drawing of The *Coya Raymi* festival by Poma de Ayala (after Garcilaso De la Vega 1961 [1604]:279]

The next part of this Chapter will consider the characteristics of the Inka Empire's economic organisation, focusing in particular on its main activity; agriculture.

3.6 The agriculturally based Inka economy

Many scholars have tried to characterise the Inka economy in the last few decades. Most of them (e.g. Prescott 1959; Baudin 1962; Bushnell 1974; Patterson 1992) followed variations of Marx's theory, calling it a communist economy. Others (for instance, Lumbreras 1974; Valdivia Carrasco 1988) have argued that it was based on Andean economy and reciprocity. According to Tamayo Herrera (1994:202), recent studies have come to the conclusion that Inka economy at the time of the *Tawantinsuyu* was based on reciprocity and redistribution, but that the whole system had nothing to do with a supposed communist economy. Within the Inka economic system there was no currency or any sort of market exchange. As already stressed, tax payment to the government was based on the *mit'a* service. Again, this kind of retribution consisted of serving in the public work force, in other words, donating labour to the government as tax. People were asked to work on road constructions, maintenance and other activities for a certain period of time. Once they had "paid" their contribution, they could return to their homes (Moseley 1992:67).

As briefly stated in the previous part, the central pivot of the Inka economy was agriculture. The Inkas cultivated various products, but in particular the ones which were most suitable for the Andean environment: indigenous tubers such as potatoes and the introduction of maize from Mexico (Guillet 1987:409). Because of the restricted distribution of fertile soils, the Inkas specialised in building agricultural terraces with remarkable irrigation systems. It has to be noticed that although these two technological innovations are regarded to be essentially Inka, they were already in use within other pre-Inka Andean civilisations (Murra 1980:19).

The solar calendar year and its division into 12 months also played a relevant part in agricultural activity, and like the religious one, there were important feasts related to it. The most important ones were: the *Chahua-Huarquiz,* a festival held during the ploughing month (July); the *Capac Siquis* in the sowing month (August); the *Uma Raymi* or the watching of the crops (October); the *Hatun Pucuy* and *Pacha Puchuy* feasts during the ripening months (February and March). Finally, the harvesting month of May concluded the Inka agricultural year (Kendall 1989:199-200).

The efficiency of the Inka Empire did not relate to the richness of the environment, but rather to the organisation of the work force (Lumbreras 1974:196). Of course, economic patterns were not similar all over the Empire. Studies carried out by Costin and Earle (1989) and Hastorf (1990) proved that in the peripheral zones of the Empire the economy did in fact change with the advent of Inka occupation, but it always maintained a local character. For instance, in the Calchaqui Valley (Argentina), instead of trying to develop the poor agricultural potential of that area, the Inkas kept and even improved the local metallurgical production of gold, silver and copper. By doing this, they took full advantage of pre-existent and well developed handicraft production systems without disturbing the economic balance of the area (Earle 1994:443).

In addition to agriculture there were many other activities within the Inka economic system. The herding of llamas was quite important. These animals provided meat, milk and wool, and their hearts were used in sacrifices. However, most importantly they were used as a major means of transport of goods. Craftsmanship was also an important part of the Inka economy, and skilled fine metal work artisans from the *Chimu* tradition occupied a quite high position in the hierarchical ladder of the mode of production (Bonavia 1991:53).

The Inka were without doubt excellent administrators, but this efficiency was only achieved after the great expansion towards the end of the 15th century AD. The next section of this chapter will consider the principal force behind this expansion: the powerful and efficient Inka Army.

3.7 The Inka military organisation

The Inka Army started to become a conquering organisation towards the end of the rule of the 9th Inka king, *Pachakuti* (Kauffmann Doig 1976:273). The climax of expansion was reached by his son, *Thupa Inka,* whereas in the last years of the Empire, under *Wayna Qhapaq*'s reign, the military policy shifted to assimilation of annexed populations and territories. Like Inka sociopolitical organisation in general, the Army also had to follow a strict hierarchical pyramid. According to Kendall (1989:98) every ten warriors formed a unit which was under control of a responsible guardian called *Chungacamayoc*. Following a decimal division the hierarchical ladder went up to the commander-in-chief, *Apusquipay*. These guardians and officers were professional soldiers, but most of the Army was formed by common people recruited into compulsory military service through the *mit'a* tax payment system. Once a military mission was over the non-professional soldiers were allowed to return to their *Ayllu* and carry on with their normal lives (Espinoza Soriano 1990:361).

As far as weaponry is concerned the Inkas mainly used clubs and spears. The most popular weapon was a star-headed mace which consisted of a club with a circular head of stone (sometimes metal) with six projecting points set in a wooden handle (see Plate 3.2).

Plate 3.2. Inka Star-headed maces from the Museum of Anthropology and Archaeology, Lima (picture: F. Menotti 1995)

After the Inkas won a battle, a series of what we would consider cruel events would take place to celebrate the victory. For instance, they used to build drums with the enemies' skin, or flutes with their bones (McIntyre 1975:55). Necklaces were made from the teeth and sometimes even the whole head was kept as a trophy (Kendall 1989:106). The Inkas had various strategies for conquest but the most effective one was without doubt persuasion, or rather effective use of the threat of destruction if the enemy did not submit (D'Altroy 1992: 75). Even though there were a few rebellions within the Empire, according to D'Altroy (1992:73), there was little imperial investment in military facilities outside the core territory of the Sacred Valley. In contrast to empires with lots of forts, to consolidate their conquests the Inkas created a well developed network of excellent roads known as the Inka highways (Earle 1994:444).

In the next section attention will be focused on this well constructed and maintained system of communication and the so-called

The Inkas: last stage of stone masonry development in the Andes

"runners" who provided one of the fastest postal services ever established in the past.

3.8 The Inka "highway" network

The Inka road system as an engineering feat was even greater than that of the Roman Empire. The roads ran along the mountainsides for about 30,000 km, and they were supplemented with retaining walls, bridges and *Tambo*, also known as inns or highway shelters (Espinoza Soriano 1990:391). The size of the roads varied between a width of 1.5 m and up to 6 m depending on the nature of the terrain. On the coast, for example, the "highways were wide and straight but unlike those in the mountains, they were not paved" (Kendall 1989:139). In various parts of the road network, suspension bridges and fords were required to cross the numerous drainages. Two of the most famous bridges of the Empire were the ones which crossed the Rio Mantaro near modern Huancayo, south of the Mantaro Valley (D'Altroy 1992:120) and the other on the Rio Apurimac on the way to Curahuasi near Cuzco (Espinoza Soriano 1990:393). The Inkas also used to build drawbridges in secluded and strategic places. A remarkable example is the one situated on the "*camino real del Inca*" near Machu Picchu (see Plate 3.3).

Especially on the main roads about every 6 to 10 km were the *Tambo* (inns) where travellers used to rest (Hyslop 1990: 274). At the end of the previous part I briefly mentioned the efficient postal service provided by the "runners" or *Chasqui*, who could only be men (Katz 1972:278). These messengers were situated every couple of kilometres, and after one had finished his distance he would pass the message to the next one and so on (Bleeker 1961:41). By this technique messages or small goods used to travel very quickly from town to town and from sea to the inland areas. It has been calculated that the distance between Lima and Cuzco (c 600 km) used to be covered in about three days (Beals 1973: 130). The maintenance of this complex road network was done by the *mit'a* tax payers. Roads and bridges were also checked by professional inspectors to make sure the communication system remained efficient. Moreover, like many other public facilities, the road network was only used by a limited number of people. Common people, for instance, were not allowed to use it (Soustelle 1994:13).

The Inka highway system is only a limited expression of the Inkas' construction skills. The latter are more fully reflected in their stone masonry.

Plate 3.3. The drawbridge near Machu Picchu (picture: F. Menotti 1995)

The Inkas: last stage of stone masonry development in the Andes

So little is known about Inka stone work constructions that one cannot avoid formulating a number of questions: How did they construct their buildings? What kinds of tools did they use? Where did they obtain their materials from? And, more intriguingly, how did they transport the heavier stone blocks?

In attempting to answer the above questions, the next chapters will mainly focus on Inka stone constructions. Especially in Chapter 4 I will be discussing all principal points regarding the Inka stone work techniques, from stone extraction to wall-fitting systems. In the last two chapters we will see how the Inkas applied their stone masonry in different constructions around the Sacred Valley and in Cuzco itself.

The Inkas: last stage of stone masonry development in the Andes

Chapter 4

Inka stone masonry

4.1 Introduction

Even though Rowe (1946) published some observations on Inka domestic and public architecture in the 1940s, a comparative examination regarding aspects of Inka stone masonry did not occur until the 1970s. By concentrating primarily on the basic structural form and characteristics of building shapes, door niches, floors and roofs, first Gasparini and Margolies (1980), then Hemming (1983), and eventually Kendall (1984, 1989), have been able to clarify and define the fundamental compounds of the general nature of Inka architecture. From these studies it has emerged that the rectangular plan with a gabled roof was a dominating feature of Inka buildings, from humble rural houses to halls and sacred temples (Gasparini and Margolies 1980:135). Occasionally curved walls were also built, but this building style, according to Hyslop (1990:7) was used in response to particular irregular terrains. Examples of these latter types of building are found in Cuzco, with the Temple of the Sun (*Qori Kancha*), the Temple of Pisaq (Urubamba Valley) and the Torreon building in Machu Picchu. An unmistakable trait of Inka architecture is its wall apertures, which were generally trapezoidal in form. They had various sizes but were placed in the walls by following specific patterns. Doors, for instance, were not wider than 1.65 m, and windows were usually around 1.25 m above ground (Kendall 1989:34). An elegant characteristic of Inka architecture was the use of double or triple-jamb doorways, of which there are many examples in Cuzco (see Plate 4.1) and in the Lake Titikaka region.

Inka stone block ornamentation was limited. Walls were rarely adorned, and when they were, it was usually in the form of small animals or geometric figures engraved directly onto the stone blocks (Bankes 1977:195; Morris and Thompson 1985 63).

Two types of Inka architecture are generally found: royal and common architecture. The latter was used by common people to build their own houses or agricultural structures, whereas the former was employed for royal constructions such as temples, priests' houses and palaces. This kind of architecture is also known as fine masonry, and it is indeed on this aspect of Inka stone constructions that my research is focused.

Plate 4.1. An Inka double-jambed door reused by the Spaniards as the entrance of the Nazarenas Church in Cuzco (picture: F. Menotti 1995)

After a brief comment on the possible origins of 15th century Inka fine stone masonry, I will consider the technical details of stone extraction, dressing and transport. Different types of wall construction are discussed towards the end of the chapter. Unlike Bushnell's (1963: 126) statement in the 1960s which said that there were only two main types of stone wall structures, we will see that recent studies have classified Inka walls into five distinct divisions, which in turn can be divided into other subcategories according to stone block shapes, wall lay-outs and profiles. In Chapters 5 and 6 we will explore how these different building techniques were employed around the Sacred Valley placing a special emphasis on Cuzco, the "navel" of the Inka Empire.

4.2 Origins of Inka stone masonry

Most of the comments written on the origins of Inka architecture and in particular its stone masonry are unfortunately purely speculative. According to Rostworowski (1988:34), Inka architecture did not have a common origin, and in spite of some similarities with the pre-Inka civilisations, the Andean stone work of the 15th century may indeed be a peculiarity of the Inka Empire. Undoubtedly, as Kendall (1976) suggests, the characteristics of the Inka stone masonry could not have been generated from nothing; the influence of previous Andean cultures is evident. The problem arises when one has to decide where the sources were. Based on her interpretation of archaeological data, Kendall (1976:94) argues that the Inkas were strongly influenced by the *Wari* (or Huari) culture. In fact, her argument is founded on the similarities between the two kinds of architectures, especially in regards to rectangular enclosures and use of a grid pattern.

Other characteristics of Inka stone masonry appear to have come from the Lake Titikaka region and in particular from Tiwanaku. The Tiwanaku culture (AD 300-1100) apparently developed an extraordinary level of stonework which shares several similarities with that of the Inkas. A number of speculations have been made regarding this matter, the most famous being that of the Spanish chronicler Cobo. He (1979 [1604]:141) argues that when *Pachakuti* (the 9th Inka ruler) went to Tiwanaku, he saw those magnificent buildings and being especially impressed by the stonework, ordered his men to copy the construction techniques because he wanted the building project in Cuzco to be of the same standard of work. Referring to Cobo's chronicle, Gasparini and Margolies (1980:7-12) maintain that the reason why there are so many similarities between Inka and Tiwanaku stone masonry is that a lot of *mitimaes* (tax labour force) from there were sent to Cuzco to help build the town. This theory is backed by the fact that the Inkas claimed to be originally from the Lake Titikaka area. Therefore, having their new heartland built using some traditional Tiwanaku techniques would have helped to reinforce their beliefs about their mythical origins. From this point arises another interesting fact for speculation. Claiming that their origins lie in the Lake Titikaka region, if that were truly the case, the Inkas might simply have brought their stonework traditions with them when they moved to Cuzco. This kind of statement is difficult to test because even with archaeological evidence, it is hard to define whether or not a specific type of construction is an invention of a particular culture. For instance, the trapezoidal form of apertures was already in use before the Inka Empire, but with some further innovations and variations the Inkas made it the hallmark of their architecture (Bonavia 1991:428). Another point of disagreement about Inka stone masonry is in relation to its chronological development.

As I have already stated in Chapter 2, the historical Inka Empire lasted less than a century and it was within that period that the Inkas developed and applied more widely different types of stone masonry. Obvious questions at this point are: were these diverse building techniques related to different periods? Or, was their diversity just a matter of function? Some scholars such as Bushnell (1963), Gasparini and Margolies (1980) and Agurto Calvo (1987) believe that variations in fine Inka masonry have no chronological implications (which coincidentally would fit with Inka concepts of time). On the other hand, Kendall (1989: 86) argues that diverse types of stone masonry developed in different periods. For instance, flat-surfaced rectangular-coursed masonry dates back to the time of *Pachakuti* (9th Inka), whereas the adoption of sunken joints and small nearly square stones was developed very late, possibly at the beginning of the 16th century. Anyway, whether or not differences in Inka masonry were determined by specific periods is not the main concern of this chapter. Instead, in the following sections, these various masonry styles will be discussed from a technical point of view, from stone extraction (quarrying) to the final construction of the Inka walls.

4.3 Inka quarries: stone block extraction

According to Rowe (1946:226) the five most important quarries used by the Inka in the Sacred Valley were: Saqsaywaman, Huaccoto, Yucay, Kachiqhata and Rumiqolqa (see Fig. 4.1).

their intended construction sites. For instance, most of the andesite used to build the finest buildings in Cuzco, such as *Acllahuasi* (the convent of the Virgins of the Sun), and *Qori Kancha* (the Temple of the Sun), was obtained from the Rumiqolqa Quarry situated about 35 km south of the town.and from the site of.

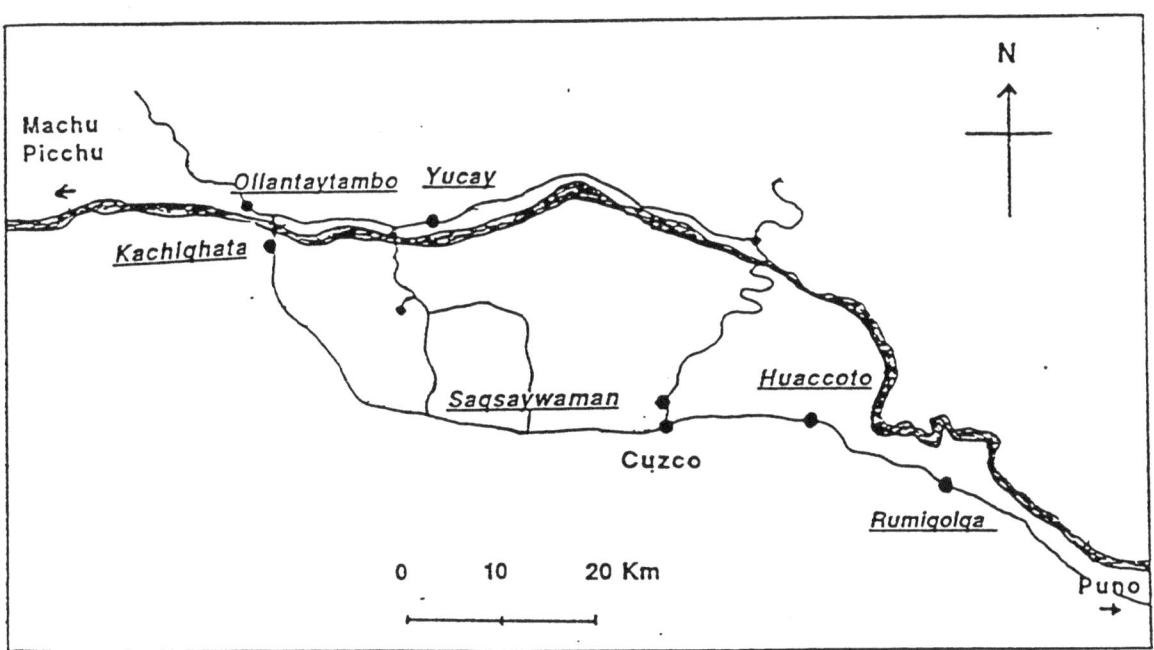

Fig. 4.1. The five most important Inka quarries in the Sacred Valley (drawing: F. Menotti 1995)

From these five quarries the Inkas extracted the material they needed according to the different types of constructions. For example, the blocks of calcite and diorite were extracted from Saqsaywaman Hill, as well as the Huaccoto Quarry. Sandstone blocks were obtained from Yucay Quarry, and andesite at Rumiqolqa and Huaccoto. Finally, the huge red rhyolite blocks used at Ollantaytambo were derived from the Kachiqhata Quarry (Agurto Calvo: 1987:120).

Because of the enormous availability of labour, the Inkas did not particularly care about the distance which separated the quarry from the building site. In fact, apart from Machu Picchu and Saqsaywaman whose construction materials were available directly from the sites themselves, most of the other sources of stone blocks were some considerable distance away from

The gigantic rhyolite blocks (some of them weighing up to 75 tons) of Ollantaytambo, 70 km north of Cuzco, , used to build the Temple of the Sun, were extracted from the Kachiqhata quarry, lying some 5 km across the Urubamba River (Protzen 1993:137). However, the Temple of the Sun was never finished perhaps owing to the civil war and the Spanish invasion (see also Chapter 2).

Of the above mentioned quarries, Rumiqolqa has always been the most studied, not only because of its long extraction history, as it was already in active use in the *Killke* period (AD 1000-1200), but also because of the clear evidence of Inka stone block extraction and dressing especially found in the Llama Pit (Protzen 1983:79) (see Fig. 4.2). In fact, some parts of this quarry are still in use.

The Inkas: last stage of stone masonry development in the Andes

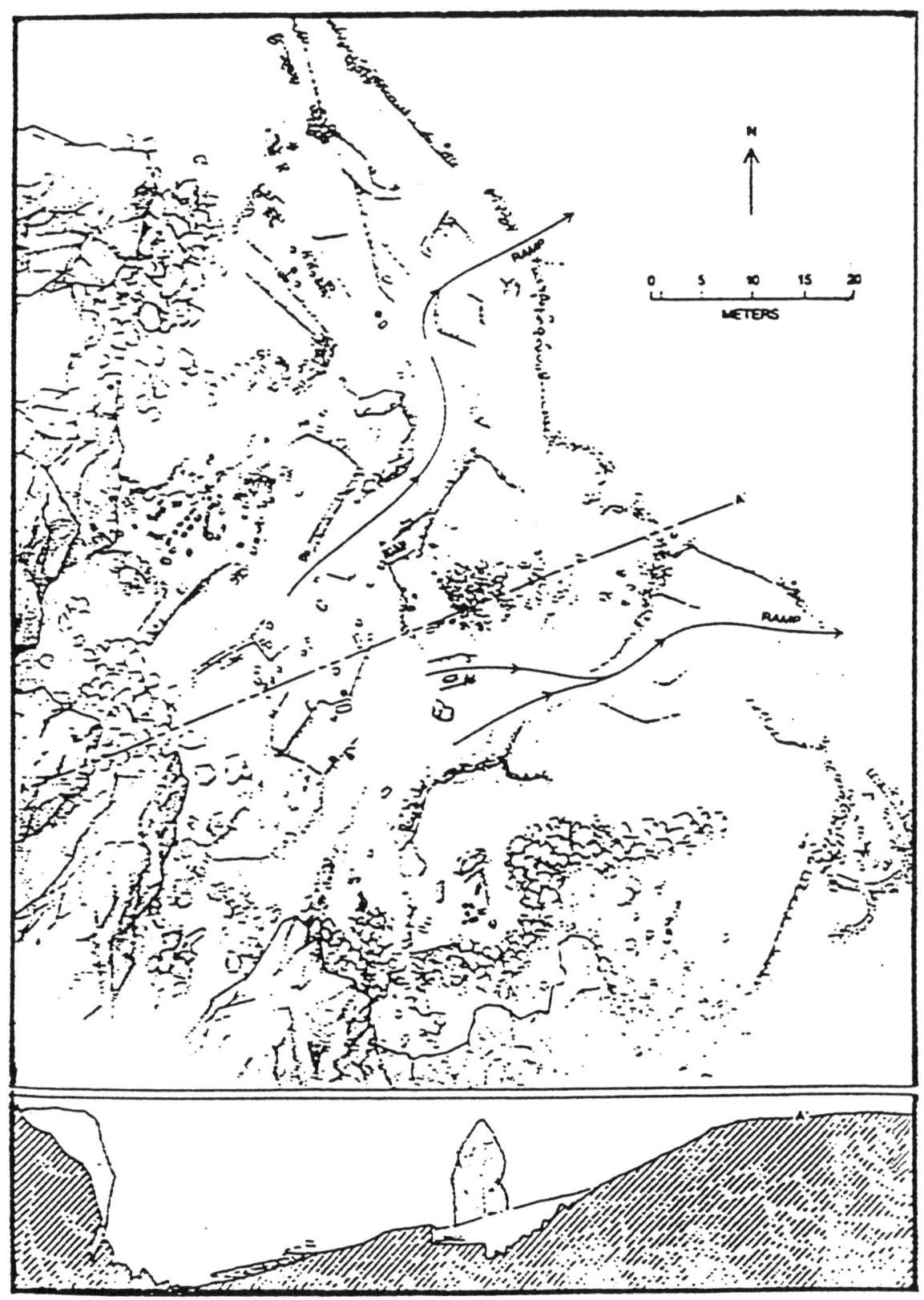

Fig. 4.2. The Llama Pit at Rumiqolqa Quarry
(after Protzen 1986:83)

The Inkas: last stage of stone masonry development in the Andes

The basic method of stone extraction was apparently very simple. Once a coarse piece of rock was fractured from the natural bedrock, the process of dressing could start. The first part to be done was the top surface (see Plate 4.2B), then the two long lateral sides (see Plate 4.2C), and finally the two other small sides (Plate 4.2D). When the blocks had assumed a parallelepipedal form, they were finely dressed and in some cases polished (see Plate 4.2E).

Plate 4.2. Different stages of stone block production at the Llama Pit of Rumiqolqa Quarry (pictures: F. Menotti 1995)

Plate 4.2A. The bedrock

The Inkas: last stage of stone masonry development in the Andes

Plate 4.2B. First stage of dressing

Plate 4.2C. Second stage of dressing

The Inkas: last stage of stone masonry development in the Andes

Plate 4.2D. Third stage of dressing

Plate 4.2E. A finished stone block

The Inkas: last stage of stone masonry development in the Andes

As we have seen, the process of stone block extraction was fairly straightforward. Understanding how the stones were cut and dressed is more complicated, and I will come back to this point later in this chapter. The next section, instead, deals with the employment of tools, the material they were made from and, of course, their provenance.

4.4 Inka stone masonry tools

A large number of scholars, such as Protzen (1983, 1986, 1993), Heffernan (1987), Agurto Calvo (1987) and Kendall (1989), agree with the conclusion that the most common tool used by the Inkas to dress stone blocks was the stone hammer (see Plate 4.3).

Protzen (1985:170) in his researches at Rumiqolqa Quarry found about 68 hammerstones scattered around the area. These stone tools are easy to recognise. They are rounded, have a smooth surface with pitted small ends and are normally of an exotic material. Most of these stones are river cobbles weighing from 200 g. to 8 kg. They are made of quarzite, rhyolite and hematite, a non-magnetic iron ore.

The hardness of both quarzite and hematite is between 5 and 7 on the Mohs scale, which corresponds to about the same hardness as that of some of the materials (e.g. basalt, rhyolite, diorite) on which hammerstones were used (see also Table 4.1).

Plate 4.3. A 2 kg stone hammer made of rhyolite found at Rumiqolqa Quarry (picture: F. Menotti 1995)

The Inkas: last stage of stone masonry development in the Andes

What makes quartzite and hematite ideal materials for hammerstones is that both are often very tough, and sometimes tougher than the rock to be worked on. Toughness, as opposed to brittleness, is defined by the resistance of the material to the propagation of incipient fractures. This level of fragmentation depends mostly on the differential cooling which occurred when the rock was originally formed, especially as far as igneous rocks are concerned (Protzen 1986:85)

blocks or detach them from bedrock. Evidence of this kind of bronze tool usage is in fact visible in two of the most used quarries situated in the Sacred Valley, the quarries of Kachiqhata and Rumiqolqa. Recent excavations at Qori Kancha Temple have also unearthed some small copper chisels, possibly employed to polish the stone blocks in the circular structure of this Temple of the Sun (W. Zanabria, pers. com. 1995).

Type of Rocks	Hardness Mohs Scale	Comp. Resist. Kg per cm^2	Brown	Yellow	White	Grey	Black	Red	Rose	Green
ANDESITE	6	1200	X			X	X		X	X
SANDSTONE	7	300-800		X	X	X		X		X
BASALT	5-6	1200				X	X			X
CALCITE	3	200-500	X	X	X	X	.		X	
QUARTZITE	5	1000	X	X	X	X			X	
DIORITE	6	1200				X				X
RHYOLITE	6-7	1200				X		X	X	X

Table 4.1. Types of rocks, hardness and compression resistance in kg per cm² (modified from Agurto Calvo 1987: 120)

Recent studies, like those of Mansur-Franchomme (1987) for instance, have brought up the possibility, already advanced by Bingham (1930), that some copper and bronze tools might have been used to dress the stone blocks. According to Heffernan (1987:9), a use-wear study of several bronze implements based on microstructural deformation, combined with inferences based on tool form, has revealed the possibility of bronze tool usage for chipping stones. As briefly mentioned in the last section, Protzen (1986:82; 1993:168) agrees with the idea that bronze tools were used; however, he argues it was only as wedges in order to separate stone

Now that we have some idea about how the Inkas extracted their stone blocks and the tools they employed, in the next section we will explore the intriguing phase of stone dressing.

4.5 Inka stone dressing techniques

There has been much speculation about how the Inkas dressed their stone blocks. The most accepted technique is the use of hammerstones, as mentioned earlier. In order to understand exactly the technique used by the Inka masons to dress the stone blocks, I will follow Protzen's experiment carried out in the 1980s. Protzen (1986:85) began his experiment by choosing three different groups of hammerstones. The first weighed between 8 and 10 kg, the second 2 to 5 kg and the last less than 1 kg. As a second step, he selected a block of andesite measuring about 25 by 25 by 30 cm. After knocking off the largest protrusions to give the stone block the shape of a parallelepiped, he started pounding the block with a 4 kg hammerstone. If the hammerstone is dropped from a height of about 30 cm, it will rebound by 15-25 cm. It can be allowed therefore to fall again without much effort. The efficiency of the strike can also be increased by giving the hammerstones a twist with the wrists just before it drops onto the surface of the stone block. By employing this sudden movement, the angle of impact is increased by up to 30-40 degrees, and consequently, also the efficiency of cutting is augmented. Once the block's six faces have been smoothed, a small hammerstone (half kilogram) must then be used to graze the edges. The business of changing the hammerstones (e.g. from big to small) should be repeated on each face of the stone block (see Fig. 4.3).

The entire process, from squaring the stone block to drafting five edges and finishing three sides, took Protzen no more than 90 minutes. What the experiment revealed was that, contrary to modern day thinking, the technique of stone dressing is very efficient and a skilled Inka mason, after years of experience, could have dressed a similar-sized stone block in less than one hour. Protzen repeated the same kind of experiment at the Kachiqhata Quarry where the huge Ollantaytambo stone blocks were quarried. Using the same procedures, Protzen (1993:173) worked out that a block of stone measuring 4.5 by 3.2 by 1.7 m would have taken 20 quarry workers about 15 days to complete. Consequently, if we assume that at Kachiqhata Quarry the labour force was say c 300 workers, the total number of 150 rhyolite blocks found next to the quarry would represent no more than eight months of work.

Physical evidence on Inka stone blocks for usage of the hammerstone pounding technique is abundant. On almost all stones of Inka walls, regardless of the type of rock, one can find scars resembling those left by the pounding on Protzen's experimental stone block.
Unfortunately, this evidence does not prove with certainty that the Inkas always used this technique, but it does at least confirm the usage of hammerstones as mentioned in some 16th and 17th centuries chronicles.

Before I consider the different types of Inka stone walls, in the next section of this chapter I will look at an aspect of Inka stone masonry which still has many questions unanswered: namely, the transport and manipulation of stone blocks from quarry to building site.

4.6 Stone block transport and manipulation

One of the most debated aspects of Inka stone masonry is the transport of stone blocks from quarry to building site. According to Baudin (1962:167), Lanning (1967:164), Outwater (1978:586), Ravines (1978:565) and Kendall (1989:163), stone blocks were transported with the help of ropes and wooden rollers. The usage of animals (llamas in particular) in stone block transport has also been well discussed. For instance, agreeing with Rowe (1946), Heffernan (1987:12) argues that llamas were employed to transport small stone blocks weighing up to 60 kg. On the other hand, Matos Mendieta (1994:94) maintains that the transport of stones was done exclusively by people (and all males) with absolutely no use of animals.

While Protzen is uncertain as to how the finely dressed stone blocks were transported from Rumiqolqa, he firmly denies any employment of wooden rollers for moving the gigantic blocks at Kachiqhata Quarry. Bengtsson (1988:6) holds the same opinion, even though she argues that the archaeological evidence for wooden implements could have easily been destroyed by the area's seasonally humid conditions. However, an examination of some of the stone blocks from Kachiqhata Quarry reveals clues which fully sustain Protzen's hypothesis. In fact, at least one side of the huge stone parallelepipeds has a smooth, yet uneven, polish traversed by fine, more or less parallel striations. Protzen (1993:177) argues that they could not be anything but drag marks, because they are not found on blocks which are still in the quarry.

The Inkas: last stage of stone masonry development in the Andes

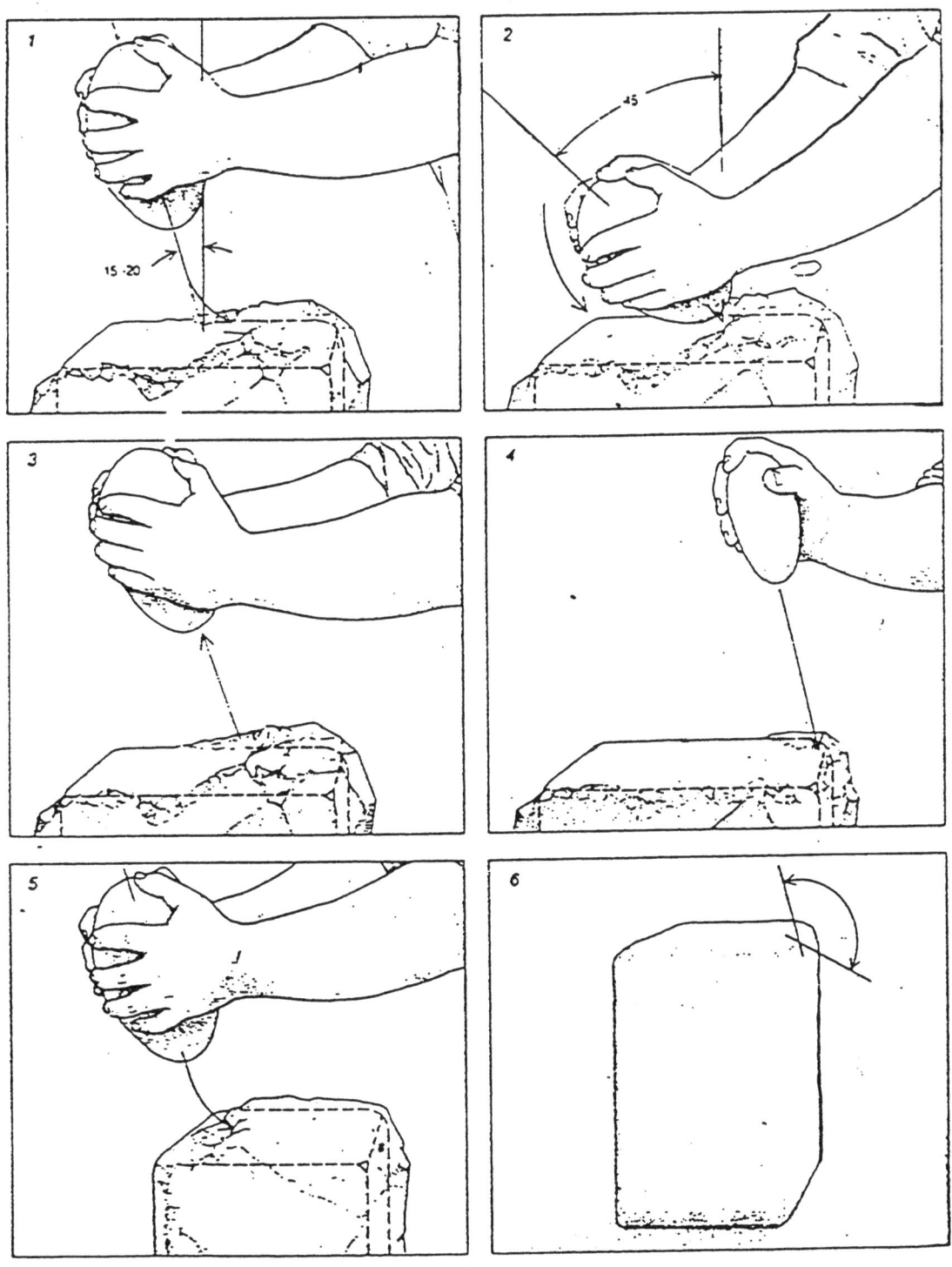

Fig. 4.3. Protzen's experiment on Inka stone dressing techniques (after Protzen 1986:84)

Even though Protzen argument is supported by the 16th and 17th centuries writers such as Cieza de Leon (1967 [1553]), Garcilaso de la Vega (1961 [1604]) and even a drawing by Guaman Poma de Ayalab (1980 [1615]; (see also Fig. 4.4), the presence of drag marks does not exclude the possibility that the stone blocks were moved in some other way, especially if one considers the large number of people needed to drag these enormous blocks, and the fact that labour was the standard and only means of taxation.

Fig. 4.4. Drawing of the transport of stone blocks by Poma de Ayala [1615] (after Espinoza Soriano 1990:210)

According to Protzen (1986:88), the force (K) required to drag any stone block depends on the coefficient of friction (f) between the stone and the material of the ramp, the weight of the block (P) and the slope of the ramp (a):

K= f . P . cos a +/- P . sin a

The + is used to compute the force needed to pull uphill and the - to drag the stone downhill. After some experiments on ballast of broken rock and compact dirt, Protzen (1993: 179) obtained a coefficient of friction of 0.75 for the former and 0.7 for the latter. Knowing that one of the largest stone blocks weighed about 110,000 kg and assuming that a person could pull 50 kg, some 2,000 people would be required to drag such a rock up an 8% gradient.

In his studies Agurto Calvo (1987:124-131), lists a number of possible techniques for transport, ranging from the employment of road beds to wooden rollers, as well as again the dragging technique with the preparation of special sliding tracks. He argues that as long as the archaeological evidence remains so limited, all the above-mentioned techniques must be considered.

Quite often scholars try to find answers that are logical within the balance between labour, time and environment, but as archaeologists we make poor palaeopsychologists. I mean, it needs to be remembered that the thinking of ancient people would often have been different to that of today. As Ponce Sanguines (1968:36) points out, a similar waste of energy to transport stone blocks also happened within the *Tiwanaku* culture long before the Inkas. Maybe, in spite of the long distance between the quarry and the building yard, the transport matter was easier than what we can imagine today, or owing to certain cosmological beliefs difficult stone block transport had other cultural implications.

Now that we know more about how the Inkas used to extract, dress and transport their stone blocks, I will consider the diverse types of stone wall construction, their subdivisions and the kinds of buildings they were employed in.

4.7 The different types of Inka stone wall construction

As briefly mentioned in the introduction to this chapter, contrary to what was believed a couple of decades ago, recent studies have revealed that there are more than two main types of stone wall construction. According to Agurto Calvo (1987:147) there are five distinct types of Inka stone wall masonry: rustic, cellular, encased, sedimentary and cyclopean. These classifications can be further subdivided according to stone block shape, wall cross-section, wall block layout, profile and texture. The adoption of these subdivisions depends on the type of wall. In fact, one wall may have them all, while another may have just one type.

The Inkas: last stage of stone masonry development in the Andes

Before looking at these five Inka wall styles, let us consider the characteristic elements of their subdivisions starting with the cross-sections.

have been classified by Agurto Calvo (1987:170) into five different forms: polygonal, pentagonal, natural, tetrangular and rectangular (see Fig. 4.6).

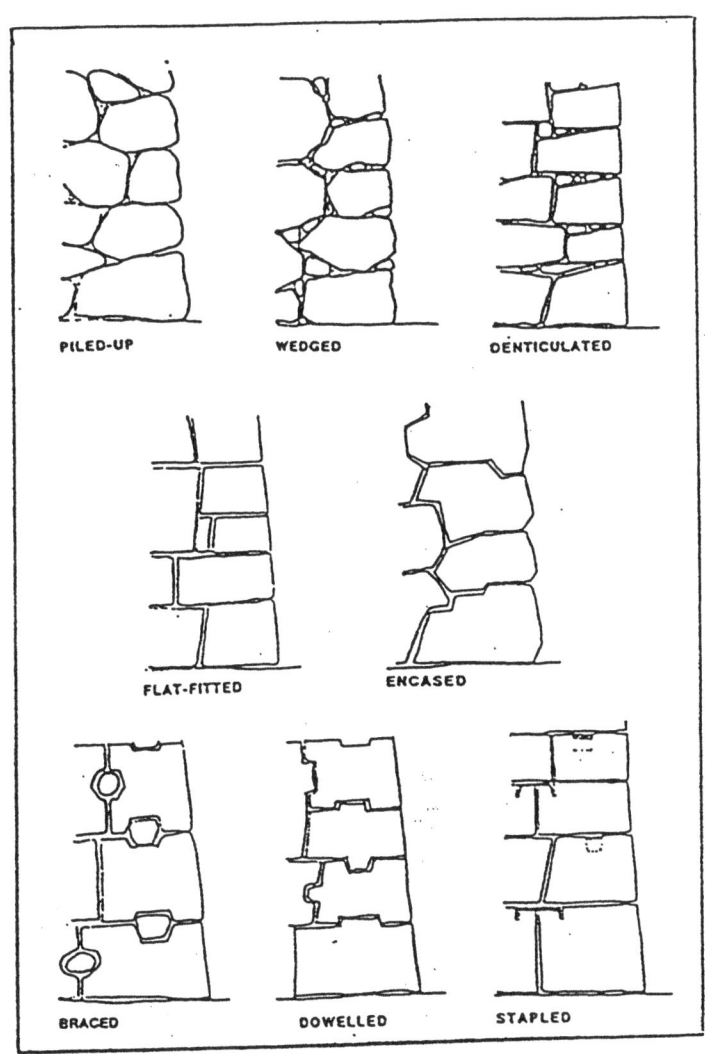

Fig. 4.5. Cross-section structures of Inka walls (after Agurto Calvo 1987:173)

Cross-sections. There are eight diverse types of cross-sections: piled-up, wedged, denticulated, flat-fitted, encased, braced, dowelled and stapled (see Fig. 4.5). In some cases these cross-sections can only be used with specific wall styles, while in some others a single style can have two or more cross-sections.

Stone block shapes. The Inkas did not have a specific standardised block shape for their stone wall constructions. According to the type and lay-out of the wall, the Inka masons adopted diverse block shapes which

The first three types (polygonal, pentagonal and natural) can be used in various wall constructions from rustic to sedimentary, whereas the last two, tetrangular and rectangular, can only be employed in the encased type of wall.

Stone block joints. Apart from the rustic type of wall, the precise joints of the stone blocks are an extremely important characteristic of Inka stone wall masonry.

The Inkas: last stage of stone masonry development in the Andes

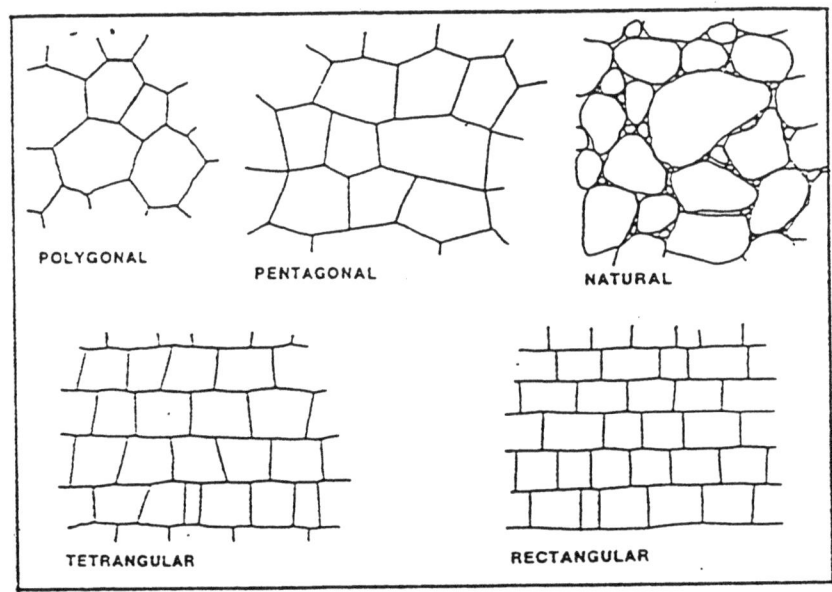

Fig. 4.6. Inka stone blocks: shape (after Agurto Calvo 1987:170)

The finest type of joint is the carved or sunken form (Heffernan 1987:13) which gives the whole wall structure a "plastic" appearance. There are three other styles of joints: natural, rustic and refined (see Fig. 4.7). Especially in royal buildings where refined joints are used, the fitting technique is so perfect that even a razor blade cannot be stuck in the joint of two blocks.

Profiles. The last elements of stone wall construction subdivisions which need to be taken into account are the profiles, which like the joints and block shape vary according to the masonry style used. Agurto Calvo (1987:171-172) has recognised six different types of profiles: natural, rough-hewed, cushioned, convex, bevelled edges and flat. These profiles also have three different kinds of textures: coarse, rough and smooth (see Fig. 4.8).

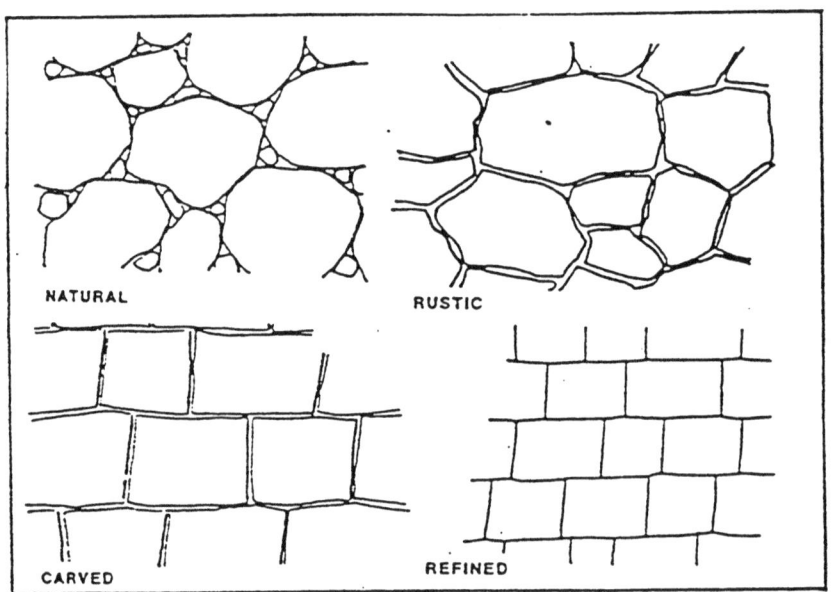

Fig. 4.7. Inka stone block joints (after Agurto Calvo 1987:171-172)

The Inkas: last stage of stone masonry development in the Andes

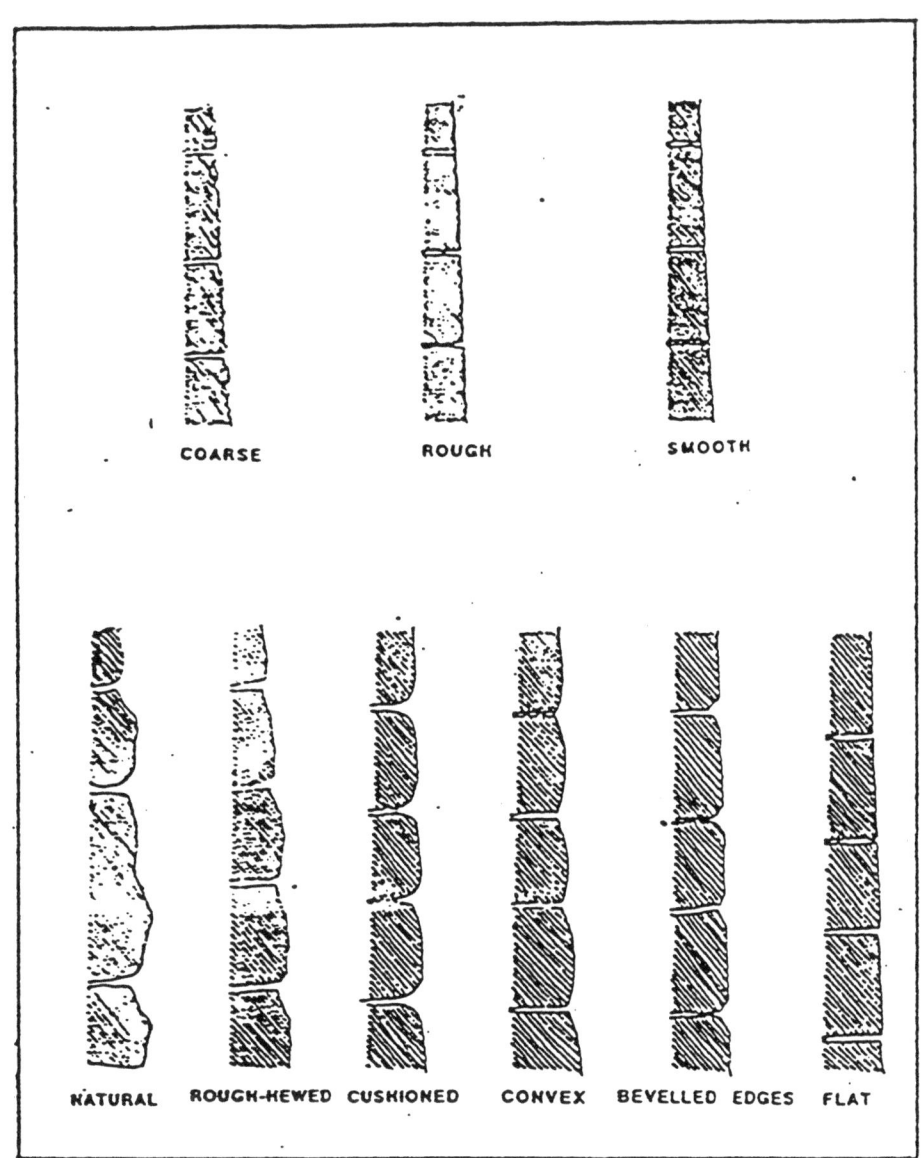

Fig. 4.8. Profiles and surface textures of Inka walls (after Agurto Calvo 1987:171-172)

The usage of diverse profiles and textures follows specific patterns. For instance, with an encased rectangular-blocked wall, it was common to employ a flat profile with a smooth texture, whereas sedimentary and cyclopean stone wall structures required convex or cushioned profiles with a fairly rough texture (W. Zanabria, pers. com. 1995).

In some cases, especially with the sedimentary type of wall, distinctive protuberances on the stone blocks can be noticed. The function of these projections is not completely clear. Protzen (1986:87, 1993:201) maintains that they may have served as points to which ropes could be attached, or as places where the force of a lever could be applied. These protuberances were apparently cut at the building site and in some cases, as Kendall (1989:165) argues, left as decorative features rather than being cut off (see Plate 4.4).

Plate 4.4. Protuberances on a encased type of Inka wall at Ollantaytambo (picture: F. Menotti 1995)

Having discussed the various subdivisions, let us now consider the five principal styles of the Inka stone wall constructions.

The rustic style (see Plate 4.5).

This style of stone wall masonry consists of a superimposition of stone blocks which have not previously been worked. The material used in this type of structure varies from place to place. In the Cuzco Valley, for example, calcite and grey diorite were used. Most of the time they were randomly collected from natural stone outcrops because the stone blocks were not worked, and because their size was never standardised. In a rustic wall either small or big stones can be found, depending on the wall structure and the shape of the stone blocks themselves (Agurto Calvo 1987:150).

The rustic style was employed for the construction of retaining walls and in particular agricultural terraces, even though it was sometimes used in so-called royal architecture if the situation required it. As far as the cross-section of this stone masonry style is concerned, it can be wedged or piled-up with a natural-coarse texture profile. Obviously, the stone block shape is natural with natural-rustic joints, which also in this case were very seldom mortared (see also Figs. 4.5 to 4.8).

The Inkas: last stage of stone masonry development in the Andes

Plate 4.5. Rustic style of stone wall construction (picture: F. Menotti 1995)

The cellular style (see Plate 4.6)

Plate 4.6. Cellular style of stone wall construction (picture: F. Menotti 1995)

The Inkas: last stage of stone masonry development in the Andes

The stone elements of the cellular style are fitted in a such a way that they come to resemble various organic cell tissues. The most common block shapes are the polygonal and pentagonal, and they are accompanied by rustic, sometimes carved joints. The profile of the wall is mainly convex, and according to the use of materials (generally calcite, andesite and sandstone) it can have either a rough or smooth texture. The cross-section is denticulated, but in some structures the encased type can also be found (Agurto Calvo 1987:152; see again Figs. 4.5 to 4.8). The cellular wall structure is mainly used for canalisations, retaining walls and agricultural terraces, and only very rarely in royal buildings.

The encased style (see Plate 4.7).

requires very hard materials such as andesite, diorite and basalt. Stone blocks in an encased wall normally have various sides, not only from a two-dimensional, but also from a three-dimensional perspective. The cross-section can therefore be difficult to determine. It can be encased, braced, dowelled and sometimes even stapled. The stone blocks do not have a standard shape, that is, it varies according to their size and the structure of the wall. The facing surfaces can be either convex or raised from carved sunken joints (Heffernan 1987:13; see also Figs. 4.5, 4.7 and 4.8). This type of Inka wall masonry was used in royal buildings around the whole sacred valley of Cuzco.

Plate 4.7. Encased style of stone wall construction (picture: F. Menotti 1995)

With this style of wall construction the stone blocks are fitted together like the pieces of a jigsaw puzzle, giving the structure a high level of stability. Because of the relevant size of the stone blocks (between 1 and 2 m^2 on the side), this style

One of the best examples is the virtually world famous twelve-sided stone block found in Hatunrumiyoc Street in Cuzco (see Plate 4.7). This form also occurs in the Temple of the Three Windows in Machu Picchu (Agurto Calvo 1987:157).

The Inkas: last stage of stone masonry development in the Andes

The sedimentary style (see Plate 4.8).

Plate 4.8. Sedimentary style of stone wall construction (picture: F. Menotti 1995)

The sedimentary style has acquired its name because the lay-out of the stone blocks resembles geological strata. The materials used in this stone masonry style are calcite, andesite and only rarely basalt. As far as the block shape is concerned, it varies from tetrangular to rectangular and sometimes even cubical, but in all cases the blocks have fine carved or refined joints. The sedimentary type of wall construction has various cross-sections, but the most common ones are encased and denticulated. Concerning the profile, it can be either cushioned, convex or with bevelled edges, but in most of the cases it has a rough texture (Agurto Calvo 1987:158; see again Figs. 4.5 to 4.8). The sedimentary type of wall can be seen throughout the Cuzco region, but the best example is the circular structure, also called the "Solar Drum", in the *Qori Kancha* Temple at Cuzco itself.

The cyclopean style (see Plate 4.9)

Because of the enormous size of the stone blocks, this wall masonry style has been named cyclopean, or "made by giants". Some stone blocks are more than 7 metres high and apart from a few modifications, their shape was kept as natural as as possible. The accentuated carved joints give the whole wall structure a "plastic" appearance, even though the convex surface is left coarse (Agurto Calvo 1987: 162; see also Figs. 4.5 to 4.8). This kind of stone wall style was employed for the construction of fortresses to protect sacred ceremonial places. A famous example is the huge zig-zag wall of the Saqsaywaman ceremonial centre, just out of Cuzco (Bruhns 1994:335).

Plate 4.9. Cyclopean style of stone wall construction (picture: F. Menotti 1995)

In this chapter we have looked closely at the different Inka techniques for stone extraction, dressing and manipulation. We have found that there are five, and not only two, diverse types of stone wall constructions, which themselves have various technical subdivisions. In the next chapter I will consider the most renowned Inka sites in the Sacred Valley of Cuzco, and we will see how the above-listed masonry styles were employed in diverse architectural constructions.

The Inkas: last stage of stone masonry development in the Andes

Chapter 5

Five types of Inka wall structure as applied to various buildings around the Sacred Valley

5.1 Introduction

In Chapter 4 I have considered the main techniques of Inka stone masonry from stone block extraction to dressing, transport and finally fitting systems. In particular we found that, contrary to what was previously thought, there are not two but five diverse types of Inka wall structure: rustic, cellular, encased, sedimentary and cyclopean. We have also clarified that these stone wall types have themselves different characteristic subdivisions depending on cross-section, stone block shape, wall block lay-out, profile and texture of the walls.

The five stone wall types and their subdivisions were applied to various kinds of buildings by following precise patterns. As pointed out in the preceding chapter, the encased and sedimentary types were essentially used for royal constructions, whereas the rustic and cellular were employed in the so-called common architecture: housing, retaining walls, irrigation and in particular agricultural terracing (Agurto Calvo 1987:49).

In this chapter I will chose six of the best known Inka sites around the Sacred Valley of Cuzco, and we will see how the various types of stone wall structure were adopted and adapted. First of all we examine Machu Picchu, the fascinating and mysterious supposedly "lost city" discovered by Hiram Bingham at the beginning of this century. Secondly, we assess Ollantaytambo, the majestic "fortress" where *Manqo Inka*, the

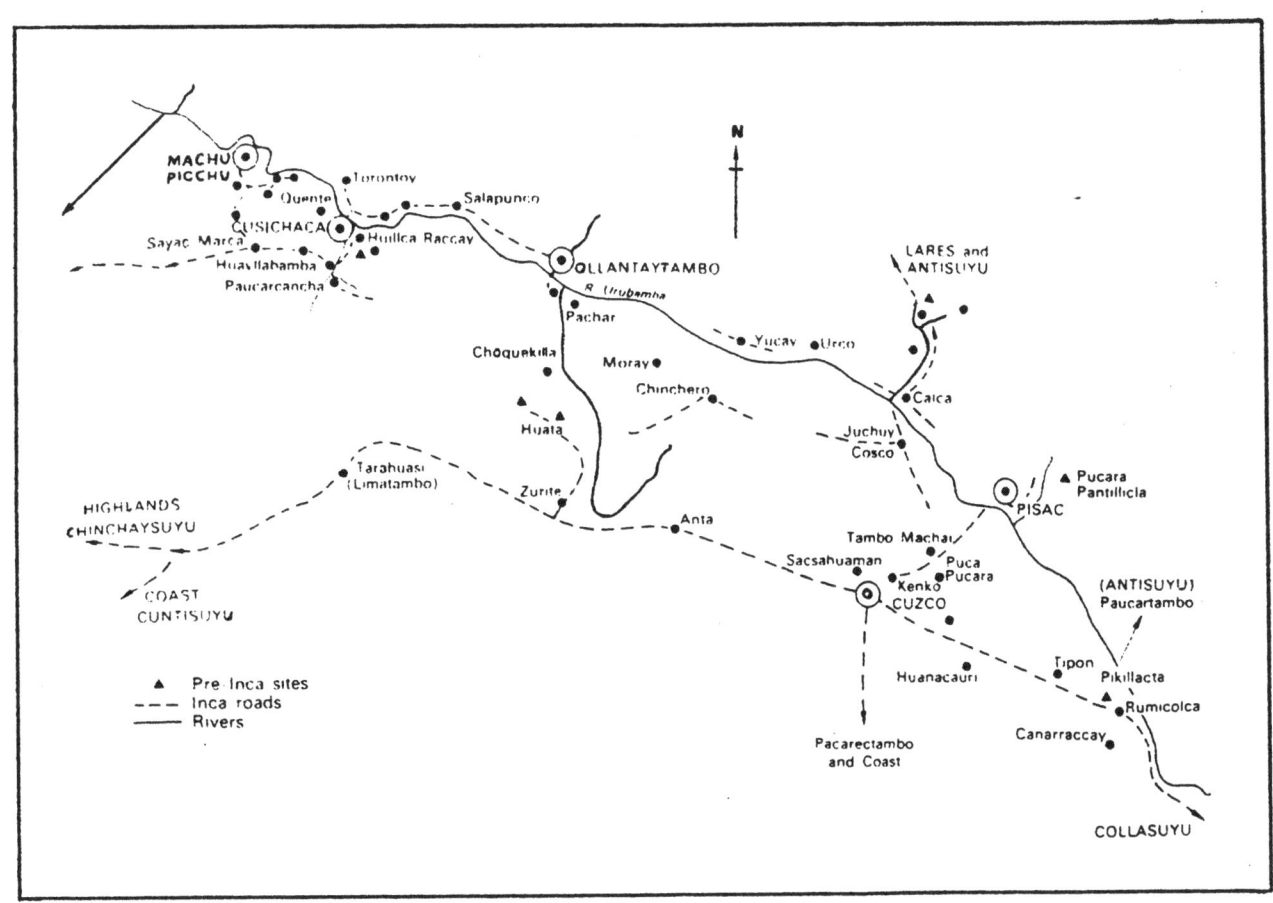

Fig. 5.1. Map of the Sacred Valley of Cuzco (after Kendall 1989:113)

founder of the Vilcabamba state, lost his last battle against Francisco Pizarro in 1536. Thirdly, we consider one of the most remarkable agricultural terrace complexes ever built by the Inkas: the Moray complex near Maras in the Urubamba Valley. The fourth place we focus on is the Inka Royal Estate of Pisaq, a ceremonial centre overlooking the Vilcanota Valley some 30 km north of Cuzco. Between Pisaq and Cuzco lies the fifth site, a small agglomerate of ruins called Tambo Machay. Because of its numerous fountains and water holes, this fifth place is also known as the "Inka Baths". Finally, the "Puma's Head", Saqsaywaman, will be examined. This remarkable "fortress" is well known for its three-tiered gigantic protection walls which enclose the ceremonial centre situated at the top of the hill (see Fig. 5.1 for the locations of these six Inka sites).

and archaeologists have often been able to do little more than speculate on its function (Espinoza Soriano 1990:334). Although Machu Picchu was known to an handful of Quechua peasants who farmed the area, the outside world was unaware of its existence until the American historian Hiram Bingham stumbled on it almost by accident on 24 July 1911 (Buse 1961:202). Bingham's search was for the lost city of Vilcabamba, the last stronghold of the Inkas. At Machu Picchu he thought he had found it.

According to Stierlin (1984:215), the building of Machu Picchu was begun by *Pachakuti* (the 9th Inka) between AD 1438 and 1471. It was probably completed by his son *Thupa Yupanki* around AD 1490. Machu Picchu in Quechua means "great peak" and in fact this *llacta* (Inka urban centre) is situated at the summit of a mountain which

5.2 Machu Picchu

Plate 5.1. Machu Picchu (picture: F. Menotti 1995)

Machu Picchu is both the best and the least known of the Inka archaeological sites. It is no mentioned in any Spanish chronicles

rises 2450 metres above sea level (Valencia Zegarra and Gibaja Oviedo 1992:5). Next to the "great peak" lies Huayna Picchu, which

The Inkas: last stage of stone masonry development in the Andes

overlooks the "citadel" of Machu Picchu like a watchtower. The impenetrable jungle, through which the River Urubamba has carved deep gorges, has helped to isolate this Inka town, which in those days, could only be reached by the ancient Inka trail. Since Bingham's relocation of the site, scholars have been speculating about the function of this fascinating Inka city. A general agreement reached, as Hyslop (1990:110) stresses, is that the religious significance may be stronger than any of the other ones. This hypothesis is reinforced by the fact that the population was very limited, maybe only an independent clergy class (Valencia Zegarra and Gibaja Oviedo 1992:321).

Machu Picchu contains various buildings displaying excellent stone work. Five of the most famous constructions within the town enclosures are: the Temple of *Pachamama* (mother earth) the *Intiwatana* (the observatory; see also Plate 3.1), the Temple compound with the temple of the Three Windows, the jail compound, the Torreon with the so-called Cave of the Royal Tomb and, of course, the remarkable agricultural complex (see Fig. 5.2 for the locations). Almost all kinds of stone wall construction can be seen within these buildings (Bingham 1979 [1930]:67). For example, there are walls of cyclopean style, such as those of the priest sacristy in the temple compound, or the more elegant encased

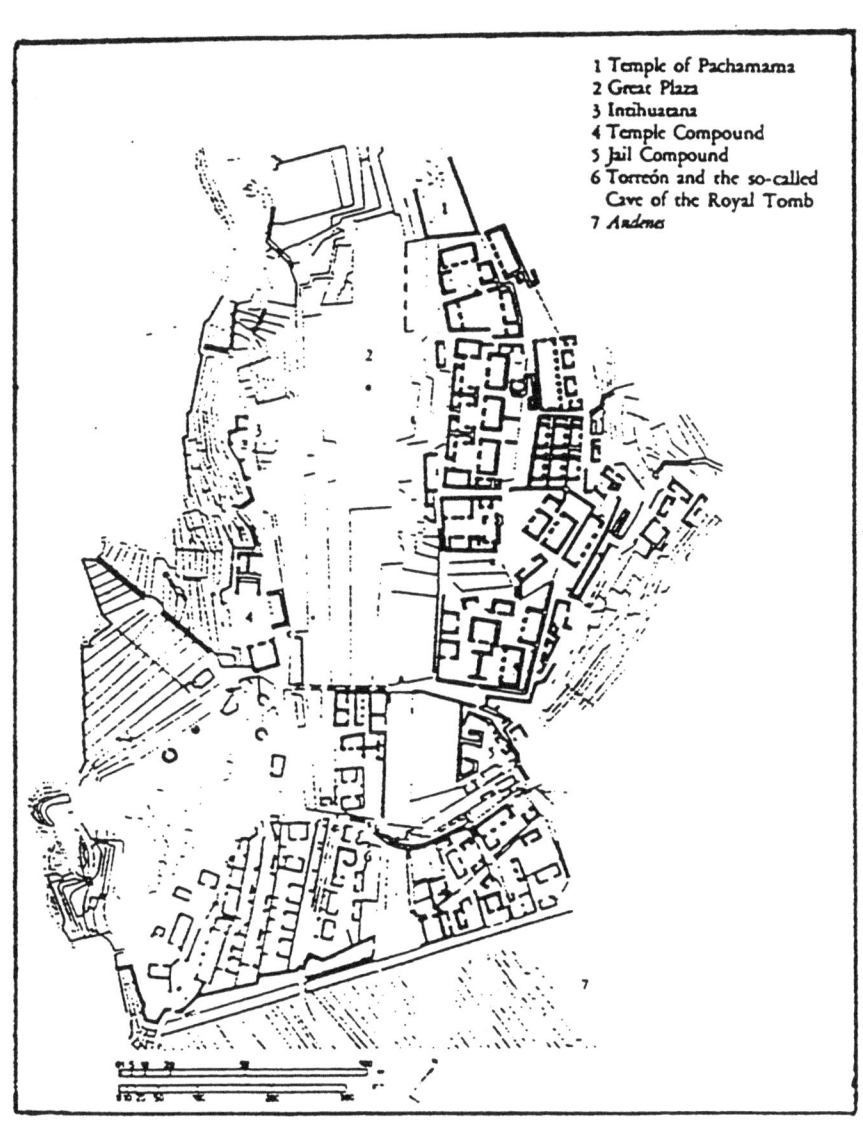

Fig. 5.2. General plan of Machu Picchu (after Stierlin 1984:212)

type used in the construction of the Temple of the Three Windows (see Plate 5.2).

of structures, thus "echoing" the characteristic shapes of the sloping roofs

Plate 5.2. The Temple of the Three Windows (picture: F. Menotti 1995)

The Torreon and the Royal Tomb below it also enclose diverse types of stone wall structure. The upper part is an excellent expression of the sedimentary style, whereas the lower part is pure bedrock sculpture combined with the encased type of stone masonry. This gives the whole construction a harmonious look of naturalism (Bingham 1979 [1930]:79; see also Plates 5.3 and 5.4).

As stated above, one of the most remarkable feats of Inka stone work and construction at Machu Picchu was the agricultural terrace complex surrounding the town. Like in other parts of the Sacred Valley, the Inkas adopted here primarily the rustic and cellular styles of wall construction to build the retaining walls of the terraces, even though the cyclopean and encased styles have also been used to increase wall stability (W. Zanabria, pers. com. 1995). One especially common building technique used at Machu Picchu was the inclination of the upper part of the wall towards the interior

(Guidoni and Magni 1977:167) and the surrounding peaks. As far as the materials used to built it, as mentioned in Chapter 4, Machu Picchu was a building yard and a quarry at the same time. In fact, all the rock used to build its edifices was directly quarried on the site. This was mainly granite.

Even though Bingham (1979 [1930]:211-215) found various copper and bronze crowbars scattered around the area, the main tools used for dressing the stone blocks were hammerstones made of diorite cobbles, presumably collected below in the Urubamba River.

The Inkas: last stage of stone masonry development in the Andes

Plate 5.3. The Torreon of Machu Picchu
(picture: F. Menotti 1995)

Plate 5.4. The Royal Tomb of Machu Picchu
(picture: F. Menotti 1995)

5.3 Ollantaytambo

Plate 5.5. The Wall of the Six "Monoliths" at Ollantaytambo (picture: F. Menotti 1995)

In the Yucay River Valley about 70 km north-west of Cuzco sits Ollantaytambo, a complex of buildings that includes a residential centre and a majestic "defensive" structure. Like Saqsaywaman, Ollantaytambo is also referred to as a "fortress". However, as far as the Inkas were concerned, the concept of fortification did not normally have a military connotation. "Fortified" places usually included ceremonial centres, royal residences and administration buildings. But, the protection sought was not always from surrounding enemies. It was meant rather as a sign of respectful isolation for those important places (Angles Vargas 1988:31).

Ollantaytambo is famous for the final defeat of *Manqo Inka* in his uprising against Francisco Pizarro in 1536. After a first defeat at Saqsaywaman *Manqo Inka* retreated to Ollantaytambo, where he initially managed to hold off the Spanish. But returning to the site with a reorganized army Pizarro forced *Manqo Inka* to seek refuge in the dense forest of Vilcabamba (Protzen 1993:19). The name Ollantaytambo simply means the Tambo (or "inn") of Ollantay. In fact, in addition to its two important sectors, the Plaza of *Manyaraki* and the religious centre, the site has a residential area where at the time of the *Tawantinsuyu* more than 1000 people could be accommodated (Angles Vargas 1988:7; see Fig. 5.3 for location).

The most impressive of the three sectors is the religious one which consists of five terraces that rise to a massive, but unfinished architectural structure at the summit: the Temple of the Sun. This sequence of the five-tiered terraces is the best expression of the different types of Inka stone masonry. The first (counting from the bottom tier), for instance, is formed by a polygonal wall built in the cellular and rustic styles.

The Inkas: last stage of stone masonry development in the Andes

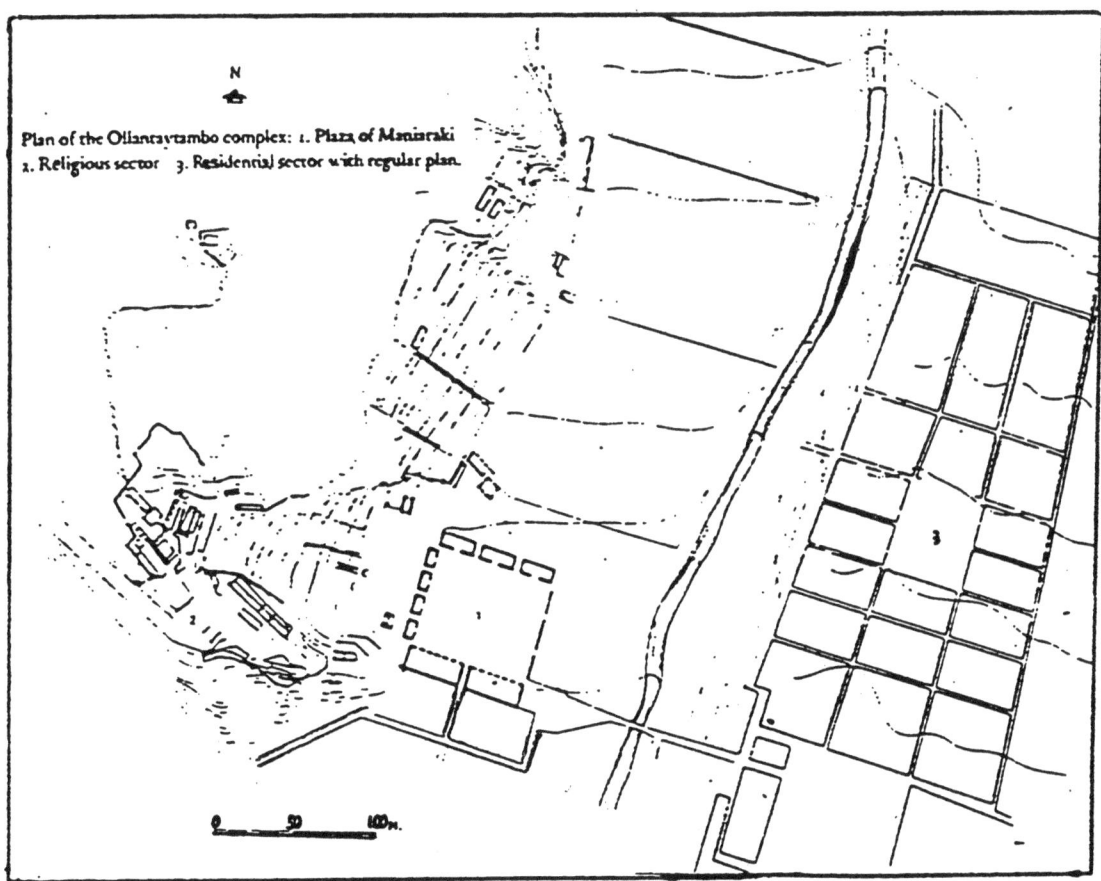

Fig. 5.3. Plan of the Ollantaytambo complex (after Gasparini and Margolies 1980:69)

The second and third tiers show a similar type of construction with the use of slightly larger components sometimes alternating with smaller ones. Approaching the top of the hill, it is in the fourth and especially the fifth tiers that the technique of dressing stone attains its most grandiose stage. A mixture of encased and sedimentary styles gives the wall a "plastic" aspect which is accentuated by the characteristic protuberances left on each block. The fifth terrace also includes two special features. One is the section of the wall with large trapezoidal niches, and the other is the majestic portal (see Plate 5.6) leading to the sacred area of the uppermost tier (Guidoni and Magni 1977:174).

At the very top of the hill lie numerous red rhyolite stone blocks, some of them semi-dressed and others ready for completion of what was perhaps to have been the most breathtaking masterpiece of Inka masonry: the Temple of the Sun.

Part of this stunning construction is the world famous six "monoliths" which make up an almost vertical wall 4.5 m high and 12 m long (Angles Vargas 1988: 36; see also my Plate 5.5). These six "monoliths" are not joined to one another directly, but have five thin slabs of stone inserted between their interstices (see Plate 5.7).

According to Protzen (1993: 85), this kind of style is entirely original. There is no other similar wall construction known throughout the entire Inka Empire. Because Ollantaytambo was built in the last two decades of the *Tawantinsuyu*, there is the possibility that a new stone masonry style was about to develop. The second hypothesis is that this "new" style was simply a peculiar characteristic created by the local Inka masons in order to distinguish the main temple from the surrounding constructions.

Plate 5.6. The portal of the sacred sector of Ollantaytambo (picture: F. Menotti 1995)

Plate 5.7. A joint of the six "monoliths" of Ollantaytambo (picture: F. Menotti 1995)

The Inkas: last stage of stone masonry development in the Andes

As already mentioned in Chapter 4, these gigantic rhyolite blocks were obtained from Kachiqhata Quarry about 6 km away across the Urubamba River. Some coarsely dressed blocks are still scattered along the path which leads to the quarry, suggesting that the construction of the temple was, suddenly and abruptly, interrupted for reasons which remain unknown, (W. Zanabria, pers. com. 1995). It may be that the civil war interrupted its completion. Finally, like many other royal estates and ceremonial centre Ollantaytambo also had its independent agricultural terrace complex situated adjacent to the five-tiered terraces of the religious sector. The retaining walls of these agricultural terraces are particularly high (2.2 m), and as usual they were built using the rustic style without the use of mortar. The material employed was not carefully selected from a quarry but more or less randomly collected from the mountain slopes nearby.

The agricultural terrace complex of Moray is one of the most remarkable feats of Inka agricultural activity. The Quechua name "Moray" denotes a territory which was occupied by the *Mullak'as* and *Misminay* tribes before the Inka Empire conquered their lands (Angles Vargas 1988:545). This agricultural complex is situated between Urubamba and Chinchero some 40 km north-west of Cuzco (see Fig. 5.1), and it consists of three major groups of concentric circles (A, B, C) plus a fourth one (D) formed by one central circle and two elevated rings. The biggest of all four groups is group A (see Plate 5.8). The diameter of the bottom circle is about 31.5 m and the whole complex consists of 12 terraces. Starting from the seventh ring, the south-eastern part of the terraces assumes a semi-circular shape as an adaptation to the mountain slope. The difference in height between the bottom and top of this amphitheatre-looking terrace group is c 69 metres.

5.4 The agricultural terraces of Moray

Plate 5.8. The agricultural terraces of Moray
(picture: F. Menotti 1995)

The Inkas: last stage of stone masonry development in the Andes

The second largest group is B which is situated 150 m north-north-west of group A; it has five concentric tiers with the bottom circle measuring about 41 m in diameter. Thirdly, we have group C which lies some 250 m north of group A. It has the same number of tiers as group B with more or less the same size. The smallest group (D) is situated c 300 m west-south-west of group A, and it consists of only a bottom circle plus two upper rings (Angles Vargas 1988:551; see also Fig. 5.4).

hammerstones were found on the site (W. Zanabria, pers. com. 1995).

Because of its double-amphitheatrical shape, it was initially thought that the Moray complex was used as a ceremonial centre. However, recent studies have confirmed that the actual function was for maize cultivation (Guidoni and Magni 1977: 174). At this point an obvious question arises: why did the Inkas build such a complex in such an isolated area?

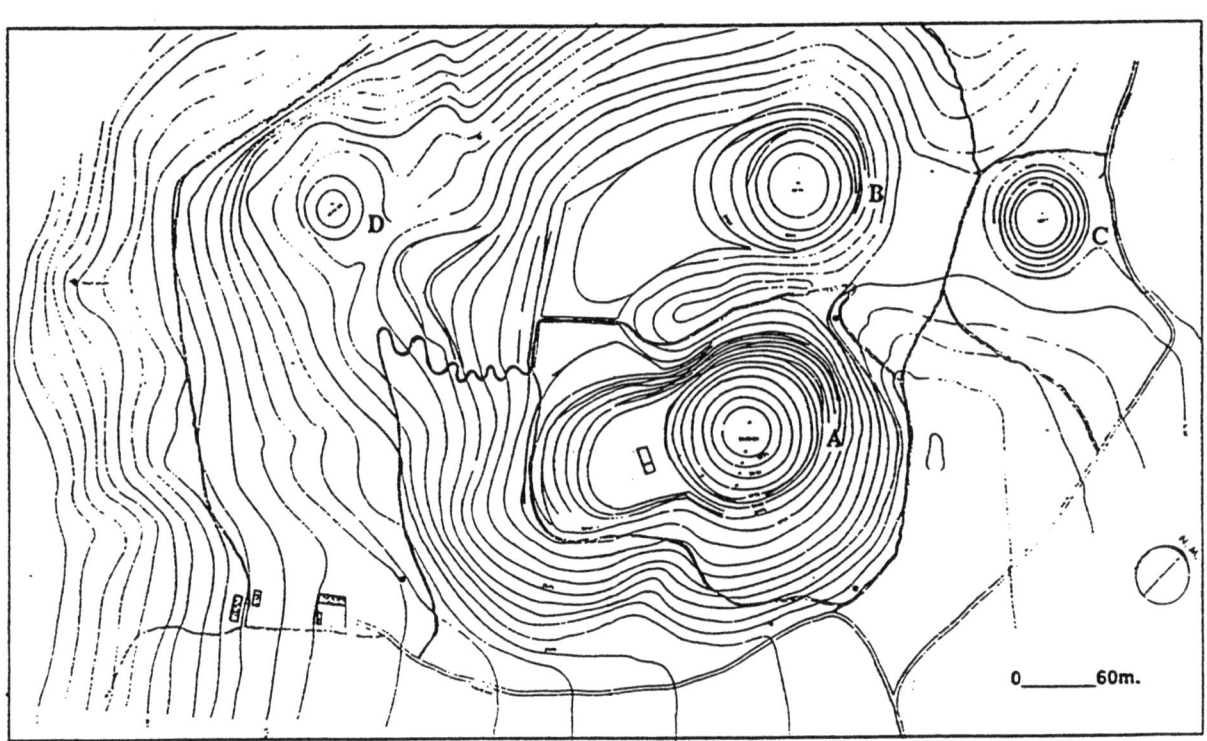

Fig. 5.4. Plan of the agricultural terrace complex of Moray (modified from Angles Vargas 1988:568-569)

The stone wall style adopted to build these concentric step-terraces was essentially the rustic with no mortar being employed. Each group of the whole complex is provided with an elaborate irrigation and drainage system (see Plate 5.9). Moreover, in order to ascend and descend the various terrace levels, ingenious diagonal flights of stairs made of flagstones were set into the terrace walls. The only material used to build this agricultural complex was andesite collected from the surrounding areas. As far as construction tools are concerned, recent excavations have revealed a lack of usage of bronze and copper tools. In fact, only

An experiment carried out by the Servicio Nacional de Meteorologia e Hidrologia or SENAMHI, (i.e. the National Institute of Meteorology and Hydrology) has proven that the temperature in the complex follows a specific pattern which is very suitable for maize cultivation. One might as well assume that the Inkas must have known about this (Angles Vargas 1988: 563-567).

Plate 5.9. Drainage system of Moray agricultural complex (picture: F. Menotti 1995)

Plate 5.10. Flights of stairs set into the terrace walls of Moray (picture: F. Menotti 1995)

5.5 The Pisaq Royal Estate

The Inka Royal Estate of Pisaq is situated some 30 km north of Cuzco. It is built on a hilltop about an hour's walk from the village of Pisaq, with a gorge on either side, Kitamayo Gorge to the west and Chongo to east (see Fig. 5.5). At Pisaq we encounter the remains of an entire city but, as Hyslop (1990:300) points out, with the characteristic features of an Inka Royal Estate. In general, Hyslop argues, the estates have extensive and fine agricultural terraces, elaborate water works, important carved and uncarved rock outcrops and numerous buildings constructed in different types of fine stone masonry. All of these characteristic features have been found at Pisaq.

A particular emphasis can be put on the elegant terraces which follow the contours of the mountain slopes (see Plate 5.12). Many of them still retain their Quechua names such as *Acchapata, Pacchayoc, Huanuhuanupata, Huinin* and *Amaru Punku* (Stierlin 1984:209). The various compounds of the whole complex are linked by narrow tracks and sometimes tunnels where the slopes of the mountains are too steep.

Like other Inka Royal Estates such as Ollantaytambo and Yucay, at Pisaq the integration of the constructions with the natural environment also seems to be a dominant factor. This also holds true for the

The Inkas: last stage of stone masonry development in the Andes

Plate 5.11. The ceremonial centre of Pisaq Royal Estate (picture: F. Menotti 1995)

Plate 5.12. The agricultural terraces of Pisaq (picture: F. Menotti 1995)

cemetery, a natural honeycombed cliff which was used to store mummified bodies. Unfortunately, out of 4000 tombs, only about 40 are still intact (in 1995), the rest having been badly looted. As I previously stated, within Inka Royal Estates we find many different kinds of buildings. According to the importance of these structures, diverse types of stone masonry were used. The whole royal complex around the ceremonial centre for instance, was built in the sedimentary style (see Plate 5.13), but with a peculiarity, the central filling of the walls. Such a design was used when the wall of an edifice needed to have two dressed or finished sides, namely on the interior and the exterior.

The finished wall consists of two parallel lay-outs of semi-dressed rectangular blocks between which a 10-15 cm gap was left. When a block layer is completed, the gap is filled with sand, pebbles or simply dirt (see Plate 5.14). The thickness of the walls depends on the size of the construction. At Pisaq the royal walls are 60-80 cm thick. Apart from cyclopean, all other stone wall construction styles were utilised at Pisaq. The encased and the sedimentary with a rough surface were employed for the main temple, whereas the rustic and the cellular were used for the agricultural terraces and the storehouses. Within the Royal complex there is also an altar sculptured in bedrock whose function has not yet been clarified.

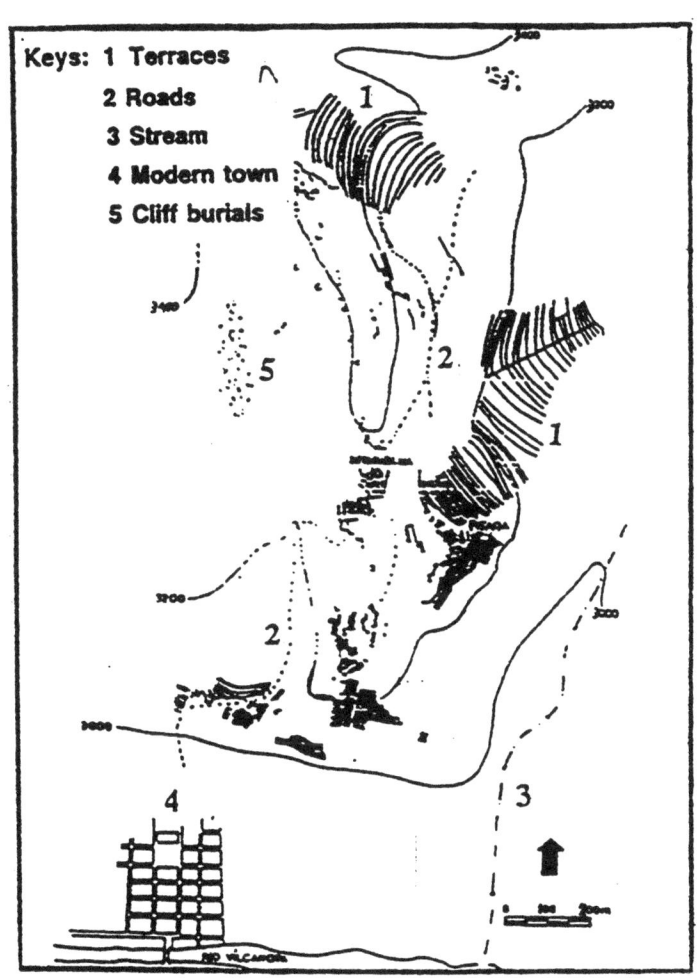

Fig. 5.5. Plan of the Pisaq complex (after Hyslop 1990:299)

Following this construction technique the rectangular stone blocks were finely dressed on five sides while the posterior one remained unworked.

Stierlin (1984:209) argues that it was a sacrificial rock, while Julio Palomino Diaz (1994: 69) maintains that it could have been an *Intiwatana* and therefore an astronomical observatory.

The Inkas: last stage of stone masonry development in the Andes

Plate 5.13. The sedimentary style of stone wall construction at Pisaq (picture: F. Menotti 1995)

Plate 5.14. Composed semi-worked stone blocks with central filling (picture: F. Menotti 1995)

The Inkas: last stage of stone masonry development in the Andes

As far as the other materials used and their provenance are concerned, there are still many uncertainties. Some authorities (e.g. Angles Vargas 1970:124) think that the sandstone might have been quarried at Yucay and the diorite may have come from Saqsaywaman. Finally, there are some mud brick constructions at Pisaq. However, some experts argue that they do not belong to the Inka Royal Estate, but were simply servants' or peasants' dwellings (W. Zanabria, pers. com. 1995).

5.6 Tambo Machay

encased and sedimentary styles, both used to build the four-tiered terraces which form the "ceremonial" baths. Then there are some retaining walls scattered around the agricultural area that were built by using the rustic type of wall structure. The *Pukara* area, thought to have been used by *Inka Thupa Yupanki* (10th Inka ruler), was mainly built in a polygonal sedimentary style, even though some encased walls can be seen there (Valencia Zegarra and Gibaja Oviedo 1990: 22).

Plate 5.15. The Baths of the Inka (Tambo Machay) (picture: F. Menotti 1995)

Tambo Machay lies 4 km out of Cuzco on the road which goes to Pisaq. This small site consists of three main sectors: the water springs, the agricultural terraces and another small area called *Pukara* (Valencia Zegarra and Gibaja Oviedo 1990:1). Tambo Machay is particularly famous for the water springs. In fact, this Inka site is also known as "the Baths of the Inka" (or simply "Inka Baths". Its water holes and fountains are very similar to those in Machu Picchu, and they probably had the same purposes: ceremonial purification as well as agricultural irrigation (Guidoni and Magni 1977:170). Within this small complex one can see three different styles of stone wall construction. Firstly, there are the

The material utilised at Tambo Machay was collected mainly around the area. However, because the quarry of Saqsaywaman was not far away, some of the big blocks employed with the encased style could have also been taken from there (W. Zanabria, pers. com. 1995).

5.7 Saqsaywaman

Saqsaywaman is one of the greatest architectural complexes ever built by the Inkas. It is situated at the top of a hill just out of Cuzco and it can easily be reached on foot by following the Inka road just after the steep Street of Resbolosa.

The Inkas: last stage of stone masonry development in the Andes

Plate 5.16. The zigzag walls of Saqsaywaman
(picture: F. Menotti 1995)

Despite its name (Saqsaywaman in *Quechua* means "imperial falcon"), this remarkable site has been considered to be the head of the puma (Cuzco was called the "Puma City" after its alleged puma shape; see also Chapter 6). According to Stierlin (1984:205) the construction of Saqsaywaman was begun by *Pachakuti* (the 9th Inka king) in the 15th century and, even though the complex was never fully completed, the final stage of actual construction was conducted by *Wayna Qhapaq* (the 11th Inka) at the beginning of the 16th century. The whole complex consists of various sectors. First of all, there are the three-tiered zigzag walls of the main fortification. Each tier is more than 6 m high and about 450 m long (Von Hagen 1963:41). At the top of the hill there is what was considered to be the ceremonial centre, which was not only protected by the aforementioned gigantic walls, but also by three majestic towers. The biggest of them was called *Muyuc Marca*. It had a diameter of 22 m, but because it was dismantled by the conquistadors after the invasion, what remains of it today is only its foundations (see Plate 5.17).

A number of scholars have speculated and debated on the matter of the function of this tower but an agreed conclusion has not been reached. Guidoni and Magni (1977:138) and Stierlin (1984:204), for example, argue that it could have been the emperor's temporary residence. Like the main one, the two lesser towers (*Sallakmarka* and *Paucamarka*) were destroyed by the Spaniards to build their colonial edifices in Cuzco (see also Fig. 5.6 for the location of the above mentioned Saqsaywaman areas).

Because of its enormous defensive walls, Saqsaywaman has virtually always been interpreted as a military fortress (Ravines 1978:571). On the other hand, the only known episode which would have turned this site into a fortress was when the Spanish stopped the revolt led by *Manqo Inka* (the founder of the Vilcabamba State), forcing him to retreat to Ollantaytambo in 1536. Until then, Hyslop (1990:57) suggests that the Saqsaywaman complex served primarily religious and ceremonial as well as storage purposes.

As previously mentioned, there are various

sectors within the Saqsaywaman complex and diverse kinds of buildings, but most of them were destroyed by the Spanish invaders. What remains to study is basically the great zigzag wall. This retaining structure was built using the cyclopean style, and it contains some of the largest stones ever moved or worked by the Inkas. The biggest is more than 6 m high, 3.5 m wide and 3 m thick (Guidoni and Magni 1977:138).

Because of the size and the natural polygonal shape of the stone blocks, the profile could only be piled-up or encased. The outside surface was left coarse, and surprisingly enough, the characteristic protuberances are absent. Finally, the material used was mainly diorite and was quarried nearby, partly from the Saqsaywaman Hill itself and partly from the Rodarero Hill situated opposite the "fortress".

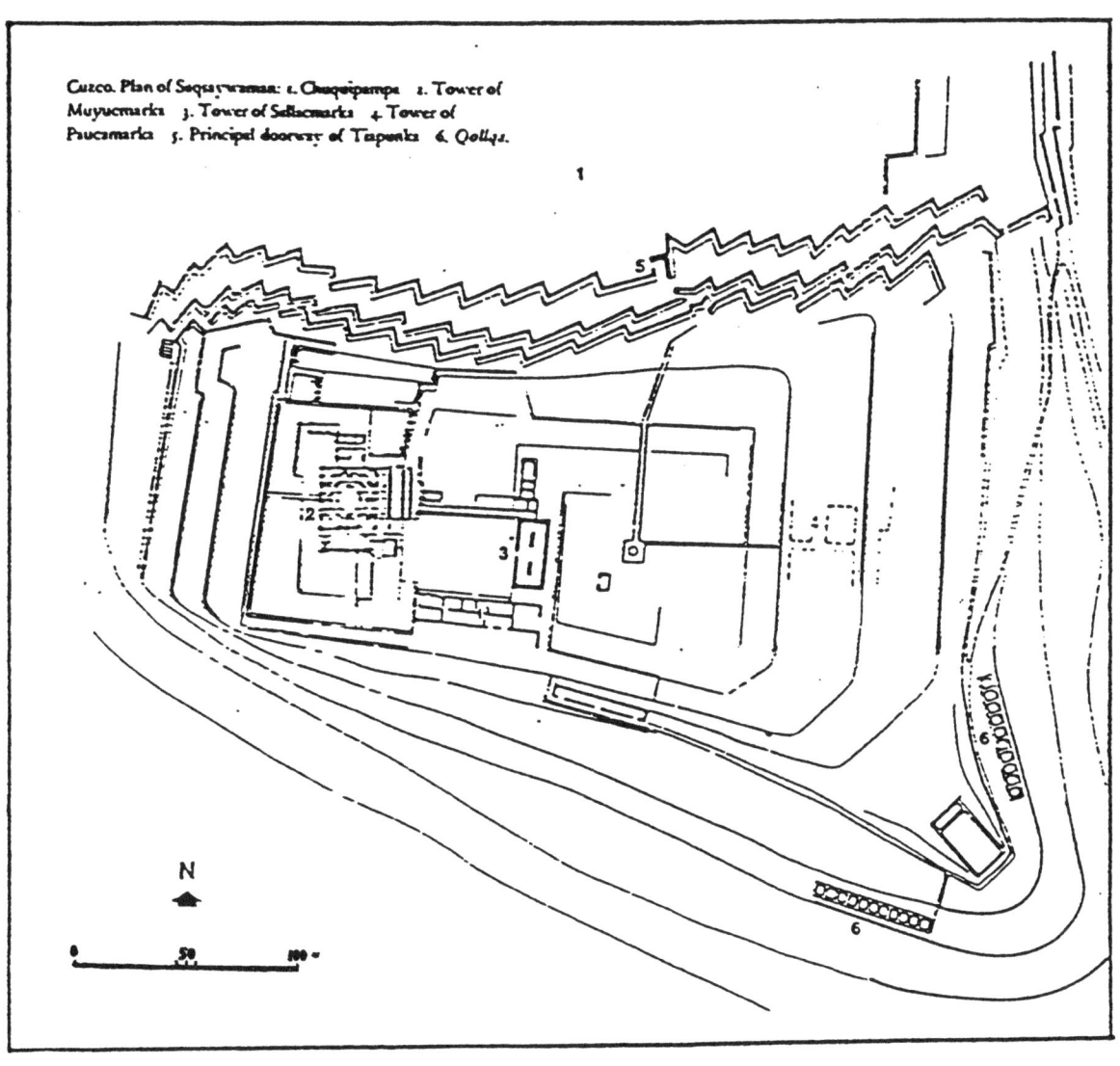

Fig. 5.6. Plan of Saqsaywaman (after Gasparini and Margolies 1980:287)

The Inkas: last stage of stone masonry development in the Andes

Plate 5.17. The foundations of the Muyuc Marca Tower of Saqsaywaman (picture: F. Menotti 1995)

In this chapter we have seen how the five different styles of stone wall construction and their subdivisions were applied to six of the most famous Inka building complexes of the Sacred Valley district. Amongst these archaeological sites I purposely left out Cuzco itself because I will focus on it in more detail in the next chapter. We will follow step by step the transformation of the Inka capital of the *Tawantinsuyu* into a Spanish colonial town, see how the Inka buildings became part of the Spanish ones, and how they in part have then "survived" until today.

The Inkas: last stage of stone masonry development in the Andes

Chapter 6

Cuzco: from Inka control to European Colonial planning

6.1 Introduction

As previously stated, Cuzco is to be discussed separately from the main Inka sites of the Sacred Valley. Having been the capital of the Inka Empire, it merits a more detailed description. Following the Inka cosmology Cuzco was regarded as the centre of the universe. In fact, in the Quechua language *cuzco* means "centre" or "navel" (Hyslop 1990:30). Throughout Chapter 6 I will consider various aspects of *Tawantinsuyu*'s main centre, focusing in particular on architecture, planning and stone masonry. After a brief section concerning the location, we follow the development of Cuzco from *Pachakuti* (9th Inka) to *Wayna Qhapaq* (11th Inka), who transformed a little village into what might have become the 15th century's largest residential agglomerate in the whole of South America.

If one looks at the prehispanic Cuzco layout, it can easily be seen that the town had a broadly feline shape. In this case, it is referred to as a puma. Some scholars believe Cuzco was initially planned with this peculiar shape in mind, whereas some others argue that the capital of the Inka Empire may be a puma but only in a metaphorical sense. In section 6.5 attention is focused on Inka walls as part of the Colonial buildings, the way in which those walls were integrated into the Spanish planning, their style of construction and their state of preservation. The final section of the chapter discusses the future perspectives and applications for Inka stone masonry technology, i.e. what "modern" people have learnt from Inka building techniques and the way they are, and will be, applied to new constructions.

6.2 Cuzco's geographical location

Cuzco is located in the south-western part of the Peruvian Andes at an altitude of about 3394 m above sea level (Diaz 1994:19).

Plate 6.1. The Plaza de Armas of Cuzco
(picture: F. Menotti 1995)

The Inkas: last stage of stone masonry development in the Andes

Because water was an important element in Inka architectural planning, the site was possibly selected as three rivers run through the city. Firstly, there is the Chunchulmayo running from west to east under the present-day Avenida del Ejercito, then the Saphy or Huatanay, flowing from northwest to southeast and passing between the two Inka plazas, the *Haukaypata* and the *Kusipata* (see Fig. 6.1).

(also called Puma Chupan or the "puma's tail"). There they form the Huatanay which, runs for c 33 km before joining the Vilcanota River (Morris and Von Hagen 1993:153). As far as the limits of the town are concerned, some scholars would argue that Cuzco was only the part between the Tullumayo and Saphy Rivers, whereas some others would include its surroundings as well.

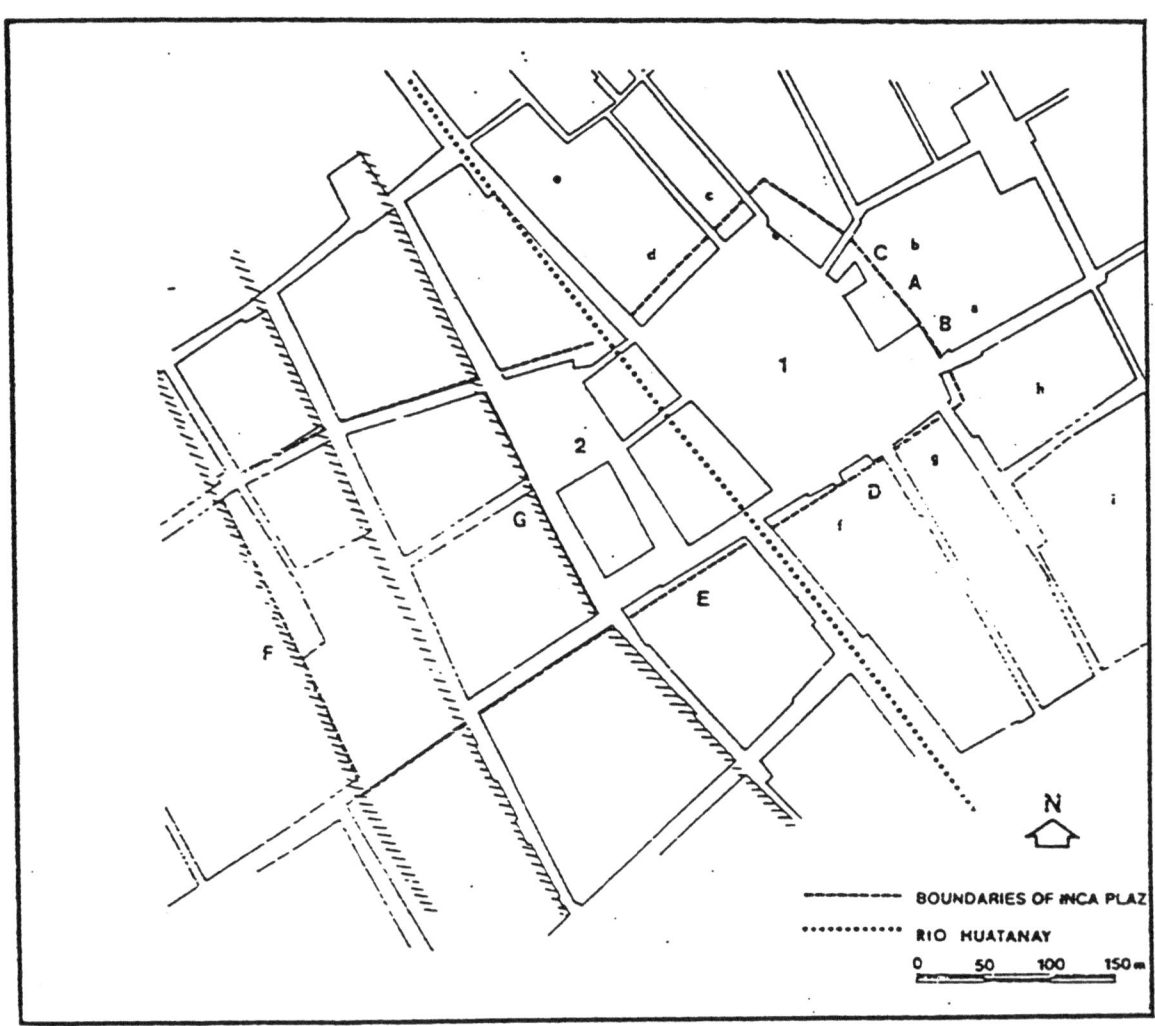

Fig. 6.1. Possible dimensions of the two main Inka squares of prehispanic Cuzco (after Gasparini and Margolies 1980:54).

Finally, a few hundreds metres east of Saphy River, and flowing in the same direction, is the Tullumayo River. The Saphy and the Tullumayo meet in the southern part of the central sector of Cuzco

Considering that Cuzco's population at the time of the *Tawantinsuyu* was about 60,000 to 80,000 people (Espinoza Soriano 1988:319), Hyslop (1990:36) has argued that the area enclosed by the two rivers (40 hectares) is too small to have contained such a number of people. Therefore, he argues, when one talks about Cuzco limits, the peripheral suburbs beyond the above two rivers must be included. Finally, even though it is situated at 13º and 29' S latitude, owing to its altitude Cuzco's climate is mild all the year round, with a relatively intense rainy season from December to March. Agriculturally speaking, the Cuzco valley is not particularly fertile, but this kind of soil is very suitable for maize and tuber cultivation.

6.3 Prehispanic Cuzco

According to Sarmiento de Gamboa (1960 [1572]: 235), the city of Cuzco was planned and rebuilt by *Pachakuti* (9th Inka). He used clay models and personally decided the location of houses, palaces and streets. For instance, only the Inkas by blood were able to live in the central sector, therefore the non-Inkas who used to live there were removed or confined to other sectors. The central part was divided in two halves. This division according to Guidoni and Magni (1977:130) had already started with the 1st Inka *Manqo Qhapaq* and was completed by Pachakuti to distinguish the people more closely related to the emperor from his more distant relatives. For example, Hanan Cuzco (the upper division) was where all Inkas by blood were located, while Hurin Cuzco (the lower division) was for the emperor's distant kin. This dual division, as Guidoni and Magni (1977:131) point out, was not a purely Inka characteristic, but it was already to be found in the pre-Inka social organisation of the so-called *Ayllu*.

Architecturally speaking, Cuzco reached its heyday towards the end of the 15th century. One of the most imposing architectural features of prehispanic Cuzco was probably not any specific building, but rather the plaza. In those days it consisted of two parts separated by the Saphy River (see Figs. 6.1 and 6.2). The eastern most part, called *Haukaypata*, was modified by the Spaniards and became today's main square, La Plaza de Armas (see Plate 6.1). West of the Saphy River was the other plaza known by the name of *Kusipata*, whose exact size has never been determined because the Spanish soon destroyed it by placing some buildings within it.

Fig. 6.2. The *Haukaypata* and *Kusipata* Squares of prehispanic Cuzco depicted by Poma de Ayala [1615] (after Hyslop 1990:38)

According to Gasparini and Margolies (1980: 53) the Kusipata Square could in fact have included modern-day Plaza San Francisco, which is situated two blocks away from the southeastern corner of Plaza de Armas. There were various important buildings around the two main squares. The best known compounds were those divided by the Street of the Sun, now called Loreto Street. On the eastern side lay *Acllahuasi* (house of the chosen women), whereas on the western side was *Amarukancha*, which according to Kendall (1989:115) was the Palace of *Wayna Qhapaq* (11th Inka; see Plates 6.2 and 6.3).

On the western side of Haukaypata Square (i.e. Plaza de Armas) stood *Qasana* Palace. Following Garcilaso de la Vega's writings, Kendall (1989:115) and Diaz (1994:40) argue that this compound was the Palace of *Pachakuti* (9th Inka). In contrast, Hyslop (1990:42) maintains that Garcilaso de la Vega's description is too vague and he attributes *Qasana* to Wayna Qhapaq (11th Inka). On the eastern part of Haukaypata Square was *Wiraqocha Inka*'s Palace, *Kiswarkancha*, which was quickly demolished by the Spaniards to make room for their cathedral.

The Inkas: last stage of stone masonry development in the Andes

Plate 6.2 The Amarukancha wall (picture: F. Menotti 1995)

Plate 6.3 The Acllahuasi wall (picture: F. Menotti 1995)

The Inkas: last stage of stone masonry development in the Andes

Next to *Acllahuasi* (on the eastern side) was *Hatukancha*, or *Pukamarka*, Palace which once belonged to *Thupa Inka* (10th Inka). Further east on the northern corner of Haukaypata Square was the palace of the 12th Inka *Washkar*. Next to *Qori Kancha*, the Temple of the Sun, was the *Kusikancha* Palace, while the *Yachahuasi*, or the School for Nobles, lay on the western side of today's Plaza de Armas, just behind *Qasana* Palace and *Kora Kora*, the residence of *Zinchi Roq'e* (2nd Inka). Not far from the ceremonial centre of Saqsaywaman were the massive retaining walls with 11 niches; these were part of the *Kolkampata* or *Manqo Qhapaq*'s Palace (Diaz 1994:34-40; see also Plate 6.4)

sacred place was *Manqo Qhapaq* (i.e. 1st Inka), who started building it exactly where the golden rod, given him by the Sun, was thrust into the earth and disappeared completely. The name *Qori Kancha* means "golden enclosure" (Gasparini and Margolies 1980:220). In fact, what shocked the Spaniards at their first glance at the place was the enormous amount of gold used to cover walls, shrines and other sacred objects. In spite of the large number of descriptions written by the Spanish chroniclers, it is difficult to determine the right size and the exact number of buildings which formed the whole *Qori Kancha* complex.

Plate 6.4. The retaining walls of *Kolkampata*, *Manqo Qhapaq*'s Palace (picture: F. Menotti 1995)

6.4 *Qori Kancha*: the Temple of the Sun

Cuzco has a considerable number of sacred places but the finest and most famous one is certainly the Temple of the Sun, known as *Qori Kancha* (Means 1973:393). Most authorities agree that the *Qori Kancha* Temple was rebuilt by *Pachakuti* in the 15th century. Some other scholars also believe that, following the legend of the Inkas' origins, the very first founder of this

According to Cieza de Leon (1967 [1572] :92 ch. 27) and subsequently confirmed by studies carried out by Gasparini and Margolies (1980), Kendall (1989), Hyslop (1990) and Diaz (1994), the *Qori Kancha* Temple consisted of four main rooms called sanctuaries and a principal building which overlooked three sides of a courtyard. The numerous shrines were adorned with precious stones, while plates of gold and silver covered the walls. In the wall structures there were different sizes of niches which were used as shelves for important or sacred objects. Sometimes these niches, if they were big enough, contained the mummies of the past rulers. Apart from acting as a ceremonial centre, another role of the *Qori Kancha* Temple was that of being the geographical and cosmological centre of the Inka *Zeque* system, which consisted of 40-41 radial lines on which approximately 328 *Huaca* (Waqa) or sacred places were located. Most of these lines were also associated with *Panaqas, Ayllu*, the two halves of Cuzco (*Hanan* and *Hurin*) and of course the four *Suyu* or quarters (*Chinchaysuyu, Antisuyu, Collasuyu* and *Contisuyu*; see also Fig. 6.3).

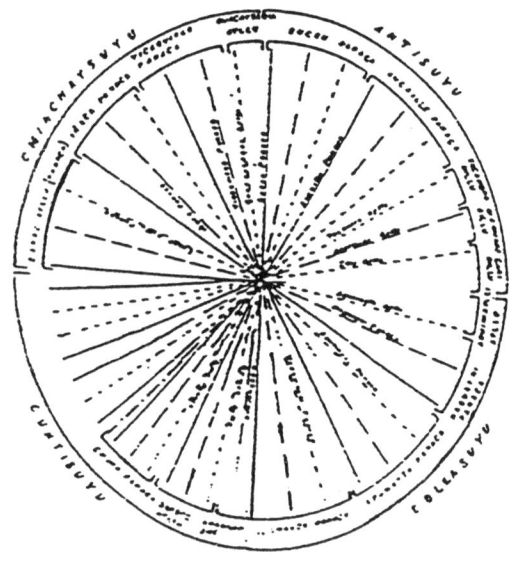

Fig. 6.3. The schematic diagram of the *Zeque* system (after Hyslop 1990:65)

Furthermore, also the ritual, political and social organisation of Cuzco seemed to have been structured round this group of imaginary lines, which according to Hyslop (1990:68) did not apparently influence the architectural planning of Cuzco City.

6.5 Cuzco: the "puma city"

As pointed out above, if we look carefully at the plan of Imperial Cuzco, we notice that it has a feline shape and in particular that of a puma (Diaz 1994: 35; see also Fig. 6.4). Most scholars who have studied the Inkas would agree with Palomino Diaz. In fact, Rowe (1967:60) had already provided this evidence in the 1960s, arguing that the area between the two rivers (the Tullumayu and Saphy), and the Saqsaywaman Hill are respectively, the body and the head of a puma. According to Palomino D. (1994:36) and Williams Leon (1994:206), although Cuzco was founded by *Manqo Qhapaq*, the actual shape of a puma was first planned by *Pachakuti* (9th Inka) and later implemented by his son *Thupa Inka Yupanki* (10th Inka). As said previously, there is some general agreement with the concept of Cuzco having a puma shape. On this notion, Gasparini and Margolies (1980:48) and Agurto Calvo (1987:104-105) have gone even further presenting two alternative puma shapes: (1) a lying or sitting puma formed by only the central part of the town, or (2) one with its legs extended which also encompassed the southwestern side of the Saphy River. Zuidema (1990), on the other hand, disagrees with most other scholars and argues that the symbolic use of a feline by the Inkas refutes the idea that the actual shape of a puma could be imposed on the capital of *Tawantinsuyu*. He maintains that the Cuzco plan, according to the *Zeque* system, has to be considered in a much broader manner. Therefore, the shape of a puma, or *leon* as the Spaniards called it, does not exist. Cuzco, Zuidema (1990:125) concludes, may only be a puma in a metaphorical sense.

6.6 Inka walls in today's Cuzco

In section 6.3 I considered the Inka buildings of Cuzco in prehispanic times. In this part I discuss how those buildings "survived" the Spanish invasion and how, at least some of them, have become part of the colonial constructions. As soon as the Spaniards took over Cuzco, the construction of Inka buildings ceased. The Colonial edifices were now planned by Spanish architects and constructed using Spanish technologies (Bennett and Bird 1960:242). While the Inka stone structures were dismantled, the material obtained was used to build Colonial structures. This process did not only happen in Cuzco, but also in other peripheral parts of the Inka Empire. Robinson (1994:63), for instance, has shown that the city of Cuenca in Ecuador was built using part of the dismantled Inka stone constructions from Tomebamba. During this destructive activity, some buildings, such as

The Inkas: last stage of stone masonry development in the Andes

the *Mayoc Marca* tower of Saqsaywaman, were completely destroyed, whereas others were used as foundations for the Colonial edifices.

western outskirts (see Plate 6.4). These retaining walls were not built in the usual style of stone construction (i.e. using natural shapes).

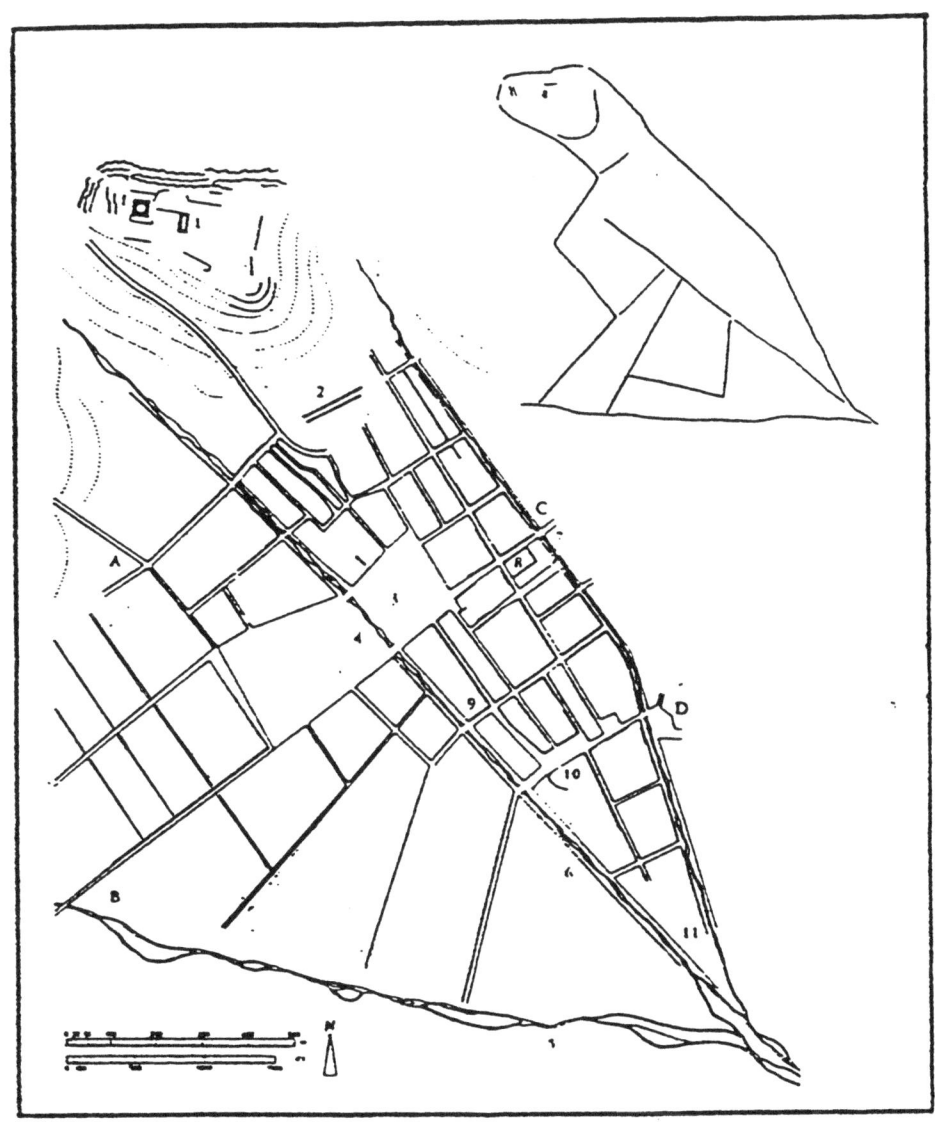

Fig. 6.4. Cuzco the puma city (after Stierlin 1984:200)

Cuzco is the best example of this integration of Inka constructions into latter Spanish ones. Walking around the town one can notice numerous Inka walls which once belonged to emperors' residences, temples and sacred places. For instance, the *Kolkampata* Palace, traditionally attributed to *Manqo Qhapaq*, forms part of the playground next to the Church of San Cristobal on the hill in Cuzco's north-

Instead, a very fine cellular style was adopted using diorite stone blocks from the nearby *Saqsaywaman* Hill. Another example of Inka walls in modern Cuzco is the *Zinchi Roq'e* Palace, the *Kora Kora*. What remains of it is visible in the courtyards of houses to the right of Calle Suecia as one walks uphill from Plaza de Armas (see Plate 6.5).

The Inkas: last stage of stone masonry development in the Andes

Plate 6.5. Part of the *Kora Kora* Palace along the Calle Suecia (picture: F. Menotti 1995)

Due to the soft material used by the Inka masons, in this case mainly sandstone quarried at Yucay, the encased and sedimentary types of wall forming the foundations of some Colonial houses are unfortunately deteriorating. Moving away from Plaza de Armas, by following Calle de Triunfo for some 300 metres, one comes to Hatunrumiyoc Street where the globally well known 12-sided stone block is situated (see Plate 4.7). This wall, as well as the one on the western side of Tullumayo Street, belonged to the *Inka Roq'e*'s Palace *Hatun Kancha*. The latter forms the foundations of the Museum of Religious Art (Wright 1984:134). These extremely well preserved sedimentary walls were again built with diorite stone blocks, quarried on the *Saqsaywaman* Hill. Heading south-east along Loreto Alley we find Inka walls on both sides. The one on the right-hand side was part of *Amarukancha* Palace, i.e. that of the 11th Inka *Wayna Qhapaq* (see Plate 6.2). This encased type of wall was built with more or less standard or same-sized rectangular blocks, whose source is unknown.

If carefully observed, it can be seen that *Amarukancha* wall has a feature rarely used by the Inkas: bas-reliefs depicting in particular serpents (Palomino D. 1994:39; see also Plate 6.6).

On the eastern side of Loreto Alley lies the longest surviving Inka wall in Cuzco. It belonged to the Acllahuasi or "House of the Chosen Women". After the Spanish Conquest the building became part of the closed convent of Santa Catalina (see Plate 6.7) and so went from housing the secluded "Virgins of the Sun" to housing the pious Catholic nuns.

The stone structure used for the *Acllahuasi* walls, along with that of the circular wall of *Qori Kancha* Temple, is the finest encased style ever adopted in the Cuzco region. The rectangular blocks of andesite were quarried and dressed at the Llama Pit of Rumiqolqa Quarry some 35 km east of Cuzco (W. Zanabria, pers. com. 1995; see also section 4.3 of Chapter 4). The best known Inka edifice of Cuzco is, without doubt, the *Qori Kancha* Temple (the Temple of the Sun), which was already described in section 6.4.

The Inkas: last stage of stone masonry development in the Andes

Plate 6.6. Bas-reliefs on the *Amarukancha* wall
(picture: F. Menotti 1995)

Plate 6.7. Part of the convent of Santa Catalina and the *Acllahuasi* Inka wall
(picture: F. Menotti 1995)

The Inkas: last stage of stone masonry development in the Andes

Part of its walls forms the foundations of the Colonial Church of Santo Domingo. Once the Inka Empire's richest temple, all that remains of *Qori Kancha* today is the stone work, or at least part of it. However, it ranks amongst the finest Inka architecture in Peru. The most famous part of the Temple of the Sun is a curved, perfectly fitted wall about 6 m in height situated just in front of the church bell tower. Inside of the Santo Domingo Church the octagonal font in the middle was once covered with solid gold. The Inka side chambers lie to either side of the courtyard, and the largest, to the right, were said to be temples dedicated to the moon and the stars (see Fig. 6.5 for a plan of the remaining Inka walls).

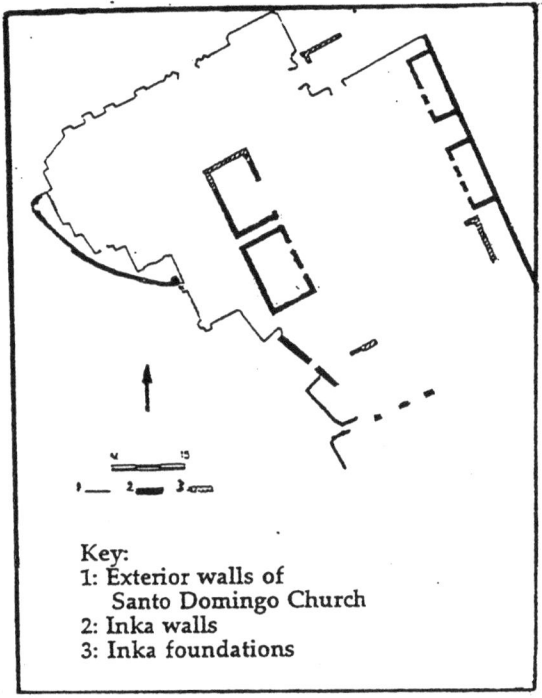

Fig. 6.5. Plan of the remaining walls of *Qori Kancha* Temple (after Hyslop 1990: 44)

The walls are perfectly tapered upwards and, with their niches and doorways, they are excellent examples of Inka trapezoidal architecture (Diaz 1994:42-44). The stone structure used is the encased-layered type, and the fitting of the individual blocks is so precise that in some places one cannot tell where one block ends and the next begins, even when one glides a finger along the joint! As for the stone blocks of the *Acllahuasi* (i.e. House of the Virgins of the Sun), the perfectly dressed andesite blocks of the *Qori Kancha* Temple were also obtained from Rumiqolqa Quarry.

In this section I have only considered the most famous of Cuzco's Inka walls. There are obviously other known and currently unknown Inka stone structures scattered around the city. Some of them are unfortunately still being dismantled, sometimes even with the knowledge of the authorities (Hyslop: 1990:30). In the next section I will focus on the future of Inka stone masonry, not so much in terms of preservation of old structures, but rather in terms of the perpetuation or renaissance of Inka stone construction technology within recent Peruvian buildings.

6.7 Inka stone masonry: perspectives for the future

Throughout Chapter 6 we have seen how Imperial Cuzco became a Colonial town. During this transformation a number of Inka edifices were dismantled, even though some of them were successfully used as foundations for Spanish buildings. At this point an obvious question arises: have "modern" people learnt anything from ancient Inka stone technology? In order to answer this question we have to take into account one of the most destructive natural forces on earth: earthquakes.

It is assumed that the reason why the Spaniards used part of Inka walls as foundations for their buildings was not because of their strength, but because of their aesthetical aspect (Agurto Calvo 1987:107). When a dreadful earthquake destroyed the whole of early Colonial Cuzco in 1650, the Spanish soon realised that the Inka walls which had been used as foundations were not damaged at all. Nevertheless, instead of understanding that it was because the edifices were so strongly built and then copying that kind of technology, the Spaniards rebuilt the town in the same way they had done before.

Three centuries later in 1950, another terrible earthquake once again destroyed Cuzco. Unfortunately, it was not until a further warning of nature (an earthquake at 5.8 on the Richter scale) in 1986, that the Peruvian authorities finally realised that the Inka walls held a secret: most of them are earthquake proof (Apacio Flores and Marmanillo Casapino 1986:11-15). Consequently, construction companies have been trying to understand and "copy" Inka building techniques and, especially as far as the sedimentary and encased styles are concerned, apply them to modern edifices. The stone blocks are obviously not as big as those the Inkas used to make, but the fitting techniques are identical. Furthermore, since rock materials have recently become more and more scarce, some building companies have been trying to make the blocks using concrete (W. Zanabria, pers. com. 1995).

After almost five centuries, only in the last few years have Peruvian people become aware of the efficiency and durability of Inka stone masonry, proving that sometimes humans do at least eventually learn from their mistakes. The Inkas' superb building techniques and what remains of their stone constructions, are no longer only a visual pleasure for interested tourists. They now provide an empirical perspective on technological "development" which will help provide a better, safer future for the next generation who will continue to live in this earthquake-prone zone of the world.

The Inkas: last stage of stone masonry development in the Andes

Chapter 7

Summary and conclusions

This work has began by giving the reader a general background on the Inkas. In Chapter 2 it has followed the development of the Inka Empire in western South America step by step throughout the succession of the Inka rulers. Despite the fact that the Inkas did not develop any sort of writing or formal calendar, say similar to the Maya of northern Central America, thanks to scholars such as Rowe (1945, 1946, 1967) archaeologists and historians have been able to calculate a broadly acceptable chronological order for both the mythological and historical times of the Inka culture.

It was shown that from *Manqo Qhapaq* (first Inka of c 1250 AD) to *Wiraqocha Inka* (8th ruler), the history of the Inkas is shrouded in mystery and mythology, but most of the legends have characters and places which can be studied empirically. With *Pachakuti* the Inka Empire begins to have a historical aspect. In spite of some divergences in opinions between the various chroniclers, the genealogies of the Inka rulers up to the Spanish invasion and later in the 16th century can now be traced. It has been established that the development of Cuzco was due to *Pachakuti* (9th Inka) during the first half of the 15th century, but that the main territorial expansion of the Inka Empire was carried out by his son *Wayna Qhapaq* after his investiture as emperor in 1471. Finally, it has been pointed out that there was already a civil war underway between the two Inka brothers *Washkar* and *Ataw Wallpa* at the time of the Spanish invasion. This fact might have facilitated Pizarro's task of taking over the *Tawantinsuyu*. In the years which followed the invasion, the Spaniards installed a number of Inka "puppets" as formal representatives of the former Inka Empire. One of them, *Manqo Inka*, rebelled against the invaders and set up the independent Inka State of Vilcabamba. The latter episode was short-lived. In effect, the powerful *Tawantinsuyu* had already disappeared.

Chapter 3 has given the reader an overview of the organisational structure of the Inka Empire. It has showed how the sociopolitical set up was typically pyramidal, with the *Sapa Inka* (the emperor) at the top and the commoners, or subject people, at the bottom. Likewise, the administrative service was strictly hierarchical. Apart from the supreme Inka, who was answerable to no-one, the two main social classes forming the administrative organisation were the *Mitimae*, or colonists, sent by the emperor to the peripheral zones of the Empire, and the local chiefs, also known as *Curaca*. An important aspect of Inka administration was law enforcement. The legal system was very strict and capital punishment was often applied.

For the outsider, particularly one from a modern nation-state, it is hard to believe that such a "perfect" administrative organisation did not developed any sort of writing system. The only recording device adopted was the *Quipu*, a series of knotted and coloured cords on which the whole administration was essentially based. This device could only be deciphered by specially trained people called *Quipucamayoc*.

A central part of Inka society was without doubt religion. The two main deities who were worship were the Sun and Wiraqocha. Many activities in Inka life were related to the religious organisation. Feasting, for instance, was the most popular one. During festivals sacrificial offerings were apparently frequent and for these sacrifices either animals such as llamas or people were used. For the latter, in particular *Aclla* or Virgins of Sun (i.e. a special group of "chosen women" picked at a young age to serve priests in their religious functions) were selected. Religion was also deeply interwoven with the sociopolitical organisation. A good example, as shown by Zuidema (1990) and Williams Leon (1994), is the *Zeque* system, a group of imaginary lines radiating from Cuzco to various Huaca (sacred place or objects) scattered around the town's vicinity.

Another important activity in Inka life was a mix of astronomy and astrology. There are various places and carved stones which are related to these activities and some of them are the so-called *Intiwatana*, which in Quechua means "where the Sun is tied up".

From an economic point of view the Inka Empire was essentially based on

The Inkas: last stage of stone masonry development in the Andes

agriculture. The main cultivated products were the tubers potatoes and the grain maize. Beans were also grown, but were of less importance. The success of Inka agricultural production was a reflection of their remarkable agricultural terraces which were provided with efficient irrigation systems.

Section 3.7 of Chapter 3 has entirely been dedicated to the Inka Army, the organisation which made possible the great and rapid expansion of the Inka Empire. The military system, like other structures of Inka society, also had a rigid hierarchical structure. The Army was formed in part of professional soldiers and in part of common people, recruited into compulsory military service through the *Mit'a* (labour) tax payment system. The efficiency and mobility of this conquering force was made possible by a well developed road network. According to Espinoza Soriano (1990) more than 30,000 km of roads ran across the entire Inka Empire. In addition, there was an efficient postal service provided by the so-called *Chasqui* (runners), who ran from post to post in relay, usually 1-2 km, carrying little parcels (e.g. of fish) or messages (e.g. *Quipu*) for the emperor. Although the highway network clearly reflects the Inkas' highly developed construction skills, these skills are more fully expressed in their stone masonry.

Chapter 4 has dealt with the main focus of this work, stone masonry. After considering the possible origins of Inka stone work methods and characteristics, it has concentrated on the entire process of stone block production. It has followed step by step the whole procedure of stone block extraction, from the time a piece of rock was detached from bedrock to the completion of a fine dressed block. The employment of tools has been discussed in section 4.3, which has emphasised the usage of bronze pry bars for the process of extraction and hammerstones for the fine dressing of the stone blocks. Section 4.4 has taken into account Protzen's (1986) experiment and showed a possible way of dressing the stone blocks, which contrary to previous thought proved to be an efficient process.

A very controversial topic of discussion has been the transport of stone blocks. Scholars' opinions on this issue were and are divided, although due to the lack of evidence to the contrary, the hypothesis advanced by Protzen (1993) seems to be the most acceptable. In fact, Protzen (1993: 175-178) argued that most of the stone materials, especially large-sized blocks, were dragged by humans along well prepared "ramps" aided by ropes. The drag marks on most of the Ollantaytambo stone blocks support his hypothesis. However, the method of transport of relatively small blocks, such as those from Rumiqolqa Quarry, has still remained a mystery.

Having clarified a number of aspects of how the Inkas extracted and dressed their stone blocks, the last section of Chapter 4 has considered the five main styles of stone wall construction: rustic, cellular, encased, sedimentary and cyclopean. These styles have been discussed in relation to their subdivision concerning block shapes, wall cross-sections, wall block lay-outs, profiles and textures.

Chapter 5 has taken into consideration six of the most famous Inka sites around the Sacred Valley of Cuzco: Machu Picchu, Ollantaytambo, the agricultural terraces of Moray, the Royal Estate of Pisaq, Tambo Machay and Saqsaywaman. It has shown how the five styles of stone wall construction described in Chapter 4 were employed in various kinds of Inka buildings.

Chapter 6 has completely been dedicated to the "centre" of the Inka Empire, Cuzco. After the Spanish invasion it underwent a dramatic transformation from Inka capital to Colonial town. Even so, as has been shown, some of great Imperial town founded by *Pachakuti* still survives today. In the final section (6.7) of Chapter 6 future perspectives and applications for Inka stone masonry have been discussed, not only from the conservation and preservation points of view, but also from technological standpoints with a view to improving modern constructions and increasing or strengthening their resistance to earthquakes.

One of the primary objectives of this research was to discuss in detail the characteristics of Inka stone masonry from an archaeological perspective. An in-depth consideration of Inka literature written by archaeologists, scholars from other disciplines and my own fieldwork in Peru, taken together, allows me to draw the following conclusions.

Five different types of stone wall construction have been recognised within Inka stone masonry. Structural features of a wall, namely cross-section, block shape, joints, profiles and textures, are used most of the time in relation to these five styles of construction. Stone dressing techniques involving the simple use of hammerstones (as demonstrated by Protzen's experiment) are not as time consuming and energy wasting as was previously thought. Although hammerstones remain the principal tools for the dressing process, recently discovered archaeological evidence shows the occasional (if not frequent at times) employment of copper and bronze bars for stone block extraction.

Some light has also been shed on one of the most intriguing aspects of Inka stone masonry, the transport of blocks from the quarry to the building site. The theory supporting the use of wooden rollers is archaeologically speaking difficult to test. On the other hand, drag marks found on big blocks (e.g. those of Ollantaytambo) show that they were apparently dragged with the help of ropes and a large number of people (about 2,000 per block). The forgoing account seems broadly reasonable, yet the practical question remains: how could 2,000 workers have been harnessed to the block so that each was contributing to the pull?

Finally, it has always been known that most of Inka walls were earthquake proof, but very little has been done to find out their structural secrets. This research will neither answer all the questions nor will it solve every mystery concerning Inka stone masonry. However, it will hopefully contribute to broadening our understanding about those remarkable walls and their builders. This knowledge can eventually be used to improve modern architectural structures in various earthquake prone regions of the world in pursuit of a more "solid" future.

The Inkas: last stage of stone masonry development in the Andes

References

AGURTO CALVO, S. (1987) *Estudio Acerca de la Construcción Arquitectura y Planeamiento Incas*. Lima: Cámara Peruana de la Construcción

ANGLES VARGAS, V. (1970) *Pisaq: Metropoli Inka*. Lima: Industrial Gráfica

ANGLES VARGAS, V. (1988) *Historia del Cusco Incaico*. Cusco: Industrial Grafica (tomo 2)

APARICIO FLORES, O. M. and MARMANILLO CASAPINO, E. (1986) *Cusco Sismo 86: Evaluacion de Inmuebles del Centro Historico*. Cusco: Municipalidad del Qosqo

BANKES, G. (1977) *Peru before Pizarro*. Oxford: Phaidon Press

BAUDIN, L. (1962) *Daily Life in Peru under the Last Incas*. New York: Macmillan

BAUER, B. S. (1992) *The Development of the Inca State*. Austin: University of Texas Press

BEALS, C. (1973) *The Incredible Incas: Yesterday and Today*. New York: Abelard-Schuman

BECK, B. (1983) *The Incas* (2nd edn.). New York: F. Watts

BENGTSSON, L. (1988) System and Variation in Building Well-fitted Stone Masonry. Amsterdam: unpublished paper presented to 46th International Congress of Americanists, 4-8, July 1988

BENNETT, W. C. and BIRD, J. B. (1960) *Andean Culture History*. New York: Lancaster Press.

BINGHAM, H. (1979) [1930] *Machu Picchu: a Citadel of the Incas*. New York: Hacker Art Books

BLEEKER, S. (1961) *Indians of the Andes*. London: Dobson

BONAVIA, D. (1991) *Peru: Hombre e Historia. De los Orígenes al siglo XV*. Lima: Edubanco

BRUHNS, K. O. (1994) *Ancient South America*. Cambridge: Cambridge University Press

BRUNDAGE, B. (1985) *Lords of Cuzco: A History and Description of the Inca People in Their Final Days*. Norman: University of Oklahoma Press.

BURLAND, C. A. (1976) *People of the Sun: The Civilization of Pre-Columbian America*. London: Weidenfeld and Nicolson

BURGER, R. L. (1989) An Overview of Peruvian Archaeology (1976-1986). *Annual Review of Anthropology* 18: 37-69

BUSE, H. (1963) *Machu Picchu*. Lima: Editorial Universitaria

BUSHNELL, G. H. S. (1963) *Peru*. London: Thames and Hudson

CIEZA DE LEON, P. (1967) [1553] *El Señorio de Los Inkas* (edited by C. Araníbar). Lima: Instituto de Estudios Peruanos

CLASSEN, C. (1993) *Inca Cosmology and Human Body*. Salt Lake City: University of Utah Press

COBO, B. (1979) [1653] *History of the Inca Empire*. Austin: University of Texas Press

COSTIN, C. L. and EARLE, T. (1989) Status Distinction and legitimation of Power as Reflected in Changing Patterns of Consumption in Late Prehispanic Peru. *American Antiquity* 54 (4): 691-714

D'ALTROY. T. N. (1992) *Provincial Power in the Inka Empire*. Washington: Smithsonian Institution Press

DIAZ, E. (1994) *Qosqo, Centro del Mundo: a 500 años de su Frustrada Desaparicion*. Qosqo: Municipalidad del Qosqo

DOBYNS, H. F. and DOUGHTY, P. L. (1976) *Peru: A Cultural History*. New York: Oxford University Press

EARLE, T. (1994) Wealth Finance in the Inka Empire: Evidence from the Calchaqui Valley, Argentina. *American Antiquity* 59 (3): 443-460

ESPINOZA SORIANO, W. (1990) *Los Incas: Economía Sociedad y Estado en la Era del Tahuantinsuyo*. La Victoria: Amaru Editores

FARRINGTON, I. S. (1992) Ritual geography, settlement patterns and the characterization of the provinces of the Inka heartland. *World Archaeology* 23 (3): 368-385

FULFORD, M.(1992) Territorial expansion and the Roman Empire. *World Archaeology* 23: 294-305

GARCILASO DE LA VEGA (1961) [1604] *The Incas: The Royal Commentaries of the Incas*. New York: Books on Demand

GASPARINI, G. and MARGOLIES, L. (1980) *Inca Architecture*. Bloomington: Indiana University Press

GROSBOLL, S. (1993) ...And he said in the time of the Ynga, they paid tribute and served the Ynga. In M. A. Malpass (ed.) *Provincial Inca: Archaeological and Ethnohistorical Assessment of the Impact of the Inca State* pp.45-76. Iowa City: University of Iowa Press

GUAMAN POMA DE AYALA, F. (1980) [1583-1615] *El primer nueva Cronica y buen Gobierno*. (edited by J. V. Murra and R. Adorno). Mexico City: Siglo XXI

GUIDONI, E. and MAGNI, F. (1977) *Monuments of Civilization: The Andes*. London: Cassell

GUILLET, D. (1987) Terracing and Irrigation in the Peruvian Highlands. *Current Anthropology* 28 (4): 409-430

HASTORF, C. A. (1990) The Effect of the Inka State on Sausa agricultural Production and Crop consumption. *American Antiquity* 55(2): 262-290

HEFFERNAN, K. (1987) Inca Stone Quarrying at Limatambo, Peru. Canberra: unpublished paper presented at the symposium: "Montage on stone", 14-15 May 1987

HEMMING, J. (1983) *The Conquest of the Incas*. Harmondsworth: Penguin

HYSLOP, J. (1990) *Inka Settlement and Planning*. Austin: University of Texas Press

JOYCE, T. A. (1969) *South American Archaeology: A re-print of T. A. Joyce's 1912 edition*. New York: Hacker Art Books.

JULIEN, C. J. (1983) *Hatunqolla: A View of Inca Rule from the Lake Titicaca*. Berkeley: University of California Press

KALLET-MARX, R. M. (1995) *Hegemony to Empire: The development of the Roman Imperium in the East from 148 to 62 BC*. Berkeley: University of California Press

KATZ, F. (1972) *The Ancient American Civilizations*. New York: Praeger Publishers

KAUFFMAN DOIG, F. (1976) *El Peru Arqueologico*. Lima: Kompactos Ediciones G. S.

KENDALL, A. (1976) Descrpcion e Inventario de las Formas Arquitectonicas Inca. *Revista del Museo Nacional* 42: 13-96

KENDALL, A. (1984) *Aspects of Inca Architecture, description, function and chronology*. Oxford: BAR (Volumes I and II, International Series 242)

KENDALL, A. (1989) *Everyday Life of the Incas*. New York: Dorset Press

LANNING, E. P. (1967) *Peru before the Incas*. Englewood Cliffs: Prentice-Hall

LUMBRERAS, L. G. (1974) *Los Origenes de la Civilización en el Perú*. Lima: Editorial Milla Batres

LYNCH, T. F. (1993) The Identification of Inca Post and Roads from Catarpe to Rio Frio, Chile. In M. A. Malpass (ed), *Provincial Inca: Archaeological and Ethnohistorical Assessment of the Impact of the Inca State*, pp. 117-144. Iowa City: University of Iowa Press

McINTYRE, L. (1975) *The Incredible Incas and their Timeless Land* Washington: National Geographic Society

MALPASS, M. A. (1993) *Provincial Inca: Archaeological and Ethnohistorical Assessment of the Impact of the Inca State*. Iowa City: University of Iowa Press

MANSUR-FRANCHOMME, M. E. (1987) *El Análisis Funcional de Artefactos Líticos*. Buenos Aires: Instituto Nacional de Antropología

MARRIN, A. (1989) *Inca and Spaniard: Pizarro and the Conquest of Peru*. New York: Macmillan

MASON J. A. (1975) *The Ancien Civilizations of Peru*. Harmondsworth: Penguin Books

MATOS MENDIETA, R. (1994) *Pumpu: Centro Administrativo Inka de la Puna de Junín*. Lima: Editorial Horizonte

MEANS, P. A. (1973) *Ancient Civilization of the Andes*. New York: Gordian Press

MILLONES, L. (1992) The time of the Inca: the Colonial Indians' quest. *Antiquity* 66: 204-216

MORRIS, C. (1988) Progress and Prospect in the Archaeology of the Inca. In Keatinge, R. W. (ed), *Peruvian Prehistory: An Overview of Pre-Inca and Inca Society*, pp. 233-256. Cambridge: Cambridge University Press

MORRIS, C. and THOMSON, D. E. (1985) *Huanuco Pampa: An Inca City and its Hinterland*. London: Thames and Hudson

MORRIS, C. and VON HAGEN, A. (1993) *The Inka Empire and Its Andean Origins*. New York: Abbeville Press

MOSELEY, M. E. (1992) *The Incas and Their Ancestors: The Archaeology of Peru*. London: Thames and Hudson

MURRA, J. V. (1980) *The Economic Organization of the Inca State*. Greenwich: JAI Press

OUTWATER, J. O. (1978) Edificación de la Fortaleza de Ollantaytambo. In Ravines, R. (ed), *Tecnología Andina*, pp. 580-589 Lima: Insituto de Estudios Peruanos

PALOMINO DIAZ, J. (1994) *Initwatanas y Numeros*. Qosqo: Municipalidad del Qosqo

PATTERSON, T. C. (1992) *The Inca Empire: the Formation and Disintegration of a Pre-Capitalist State*. New York: Berg Publishers

PONCE SANGINES, C. (1968) *Las Andesita de Tiwanako*. La Paz: Editorial Los Amigos del Libro

PRESCOTT, W. H. (1959) *History of the Conquest of Peru: with a preliminiary view of the civilisation of the Incas*. London: George Allan and Unwin

PRESCOTT, W. H. (1970) *The World of the Incas*. New York: Tudor

PROTZEN, J.-P. (1983) Técnicas de la Cantería Inca. *Revista del Museo Nacional* 23: 76-84

PROTZEN, J.-P. (1985) Inca quarrying and Stonecutting. *Journal of the Society of Architectural Historians* 44 (2): 161-182

PROTZEN, J.-P. (1986) Inca stonemasonry. *Scientific American* 254: 80-88

PROTZEN, J.-P. (1993) *Inca Architecture and Construction at Ollantaytambo*. Oxford: Oxford University Press

RAVINES, R. (1978) Edificación y Vivienda. In Ravines, R. (ed), *Tecnología Andina,* pp. 557-571. Lima: Instituto de Estudios Peruanos

REICHE, M. (1989) [1968] *Secreto de la Pampa*. Lima: Editorial e Imprenta Enotria

ROBINSON, R. (1994) The Palace of the Inca at Tomebamba. *History Today* 44(1): 62-63

ROSTWOROWSKI DE DIEZ CANSECO, M. (1988) *Historia del Tahuantinsuyu.* Lima: Instituto de Estudios Peruanos.

ROWE, J. H. (1945) Absolute Chronology in the Andean Area. *American Antiquity* 10 (3): 265-284

ROWE, J. H. (1946) Inca Culture at the Time of the Spanish Conquest. In Steward, J. (ed), *Handbook of South American Indians,* pp. 183-330. Washington: Smithsonian Institution

ROWE, J. H. (1967) What kind of City was Inca Cuzco? *Ñawpa Pacha* 5: 59-76

ROWE, J. H. (1970) La Arqueología del Cuzco como Historia Cultural. In Ravines, R. (ed), *100 Años de Arqueología en el Peru,* pp.549-563. Lima: Instituto de Estudios Peruanos

SARMIENTO DE GAMBOA, P. (1960) [1572] *Historia de los Incas*. Biblioteca de Autores Españoles. Madrid: Ediciones Atlas tomo 135, pp. 193-279

SCHREIBER, K. J. (1993) The Inca Occupation of the Province of Andamarca Lucanas, Peru. In Malpass, M. A. (ed), *Provincial Inca: Archaeological and Ethnohistorical Assessment of the Impact of the Inca State,* pp. 77-116. Iowa City: University of Iowa Press

SMITH, M. E. and BERDAN, F. F. (1992) Archaeology and the Aztec Empire. *World Archaeology* 23: 353-367

SOUSTELLE, J. (1994) *The Land of the Incas*. London: Thames and Hudson

STIERLIN, H. (1984) *Art of The Incas and its Origins*. New York: Rizzoli

TAMAYO HERRERA, J. (1994) Los Incas y el Cuzco: Arqueología del Cuzco. In Curatola, M. and Silva-Santisteban, F. (eds), *Historia y Cultura del Peru*, pp. 195-206. Lima: Departamento de Impresiones de la Universidad de Lima

TELLO, J. C. (1937) La Civilización de los Incas. *Letras* 6: 5-37

TOPIC, J. R. and LANGE TOPIC, T. (1993) A Summary of the Inca Occupation of Huamachuco. In Malpass, M. A. (ed), *Provincial Inca: Archaeological and Ethnohistorical Assessment of the Impact of the Inca State,* pp. 117-144. Iowa City: University of Iowa Press

URTON, G. (1990) *The History of a Myth: Pacariqtambo and the Origins of the Inkas*. Austin: University of Texas Press

VALDIVIA CARRASCO, J. (1988) *El Imperio Esclavista de los Inkas*. Lima: Editorial e Imprenta Desa

VALENCIA ZEGARRA, A. (1979) *Colección Arqueológica Cusco de Max Uhle*. Cusco: Instituto Nacional de Cultura

VALENCIA ZEGARRA, A. and GIBAJA OVIEDO, A. (1990) *Excavaciones y puesta en Valor de Tambo Machay*. Cusco: Instituto Nacional de Cultura

VALENCIA ZEGARRA, A. and GIBAJA OVIEDO, A. (1992) *Machu Picchu: Investigación y Conservación del Monumento Arqueológico después de Hiram Bingham*. Qosqo: Municipalidad del Qosqo

VON HAGEN, V. W. (1963) *The Incas: People of the Sun*. Leicester: Brockhampton

VON HANSTEIN, O. (1971) [1925] *The World of the Incas*. Freeport: Books for Libraries Press

WILLIAMS LEON, C. (1992) *Sukankas y Ceques: La Medición del Tiempo en el Tahuantinsuyo*. Lima: Pachacamac Museo de La Nación

WILLIAMS LEON, C. (1994) Inca Arquitectura y Urbanismo. In Curatola M. and Silva-Santistebán (eds), *Historia y Cultura del Peru*, pp. 195-208. Lima: Departamento de Impresiones de la Universidad de Lima

WRIGHT, R. (1984) *Cut Stones and Crossroads: a Journey in the Two Worlds of Peru*. New York: The Viking Press

ZIOLKOWSKI, M. S. and SADOWSKI, R. M. (1989) *Time and Calendar in the Inca Empire*. Oxford: BAR (International Series 479)

ZUIDEMA, R. T. (1964) *The Ceque System of Cuzco: The Social Organization of the Capital of the Inca*. Leiden: Brill

ZUIDEMA, R. T. (1982) Myth and History in Ancient Peru. In Rossi, I. (ed), *The Logic of Culture: Advances in Structural Theory and Methods*, pp. 150-175. South Hadley: Brill

ZUIDEMA, R. T. (1983) Hierarchy and Space in Incaic Social Organization. *Ethnohistory* 30 (2): 49-75

ZUIDEMA, R. T. (1990) *Inca Civilization in Cuzco*. Austin: University of Texas Press.

www.ingramcontent.com/pod-product-compliance
Lightning Source LLC
Chambersburg PA
CBHW061546010526
44113CB00023B/2812